Science Fair Handbook

The Complete Guide for Teachers and Parents

Anthony D. Fredericks
and Isaac Asimov

Illustrations by
Phyllis Disher Fredericks

A GOOD YEAR BOOK™

GOOD YEAR BOOKS
Tucson, Arizona

To Walt Dudley – for his warm friendship, and his work with the Pacific
Tsunami Museum in preserving the stories of the past while educating a new
generation.

—A.D.F.

Good Year Books
are available for most basic curriculum subjects plus many enrichment areas.
For more Good Year Books, contact your local bookseller or educational dealer.
For a complete catalog with information about all Good Year Books, please contact:

Good Year Books
PO Box 91858
Tucson, AZ 85752-1858
www.goodyearbooks.com

Book Design: Elaine Lopez
Cover Design: Dan Miedaner
Design Manager: M. Jane Heelan
Editor: Leslie Hann
Executive Editor: Judith Adams

ISBN 1-59647-029-1
Previous ISBN 0-673-59964-7

GOOD YEAR BOOKS

2 3 4 5 6 7 8 9 ML 09 08 07 06 05 04

Contents

Acknowledgments

Authors, like scientists, are dependent on others to guide their investigations, support their creativity, and sustain their endeavors. Several people were involved in the shape and direction of this book.

To Chris Jennison, who brought the two authors together for the first edition of this book, an eternal note of gratitude is extended. Chris is an "author's editor," one who nurtures and supports writers in countless ways from initial concept to published book.

Three educators who unselfishly contributed their ideas and materials deserve special thanks and appreciation: Mike Misko of the Catasauqua (PA.) Area School District, Jim McKinney of the Southern York County (PA.) School District, and Nat Harmon of the Dallastown (PA.) Area School District.

A special note of gratitude is extended to the scores of teachers from Hawaii to Maine who contributed their ideas in telephone conversations and E-mail messages. A resounding cheer is offered to the dozens of teachers around the United States who opened their classroom doors to share the joy, magic, and excitement of science fairs.

Most important, great thanks and appreciation are extended to the many students who were interviewed and observed throughout the creation of this book.

Finally, it is hoped that a new generation of young scientists will be the ultimate benefactors of the energy and dedication of the teachers, students, and colleagues who contributed to this book. That will be the finest acknowledgment of all!

About the Revised Edition

When *The Complete Science Fair Handbook* was first published in 1990, we were not prepared for the overwhelming and enthusiastic response it received from thousands of teachers across the country. Teachers from urban, suburban, and rural schools used the book to share the excitement of science fairs with their students. Conversations with educators in school districts all over the United States indicated to us that a rich vein had been tapped—one that could invigorate both the teaching and learning of science.

Thousands of teachers confirmed our belief that fun, exciting, and easy-to-do science fairs could enhance an inquiry-based approach to science education.

Since the book was first published, science education has changed dramatically. The new National Science Education Standards, mind-boggling new Web sites, and a reconfiguration of science curricula across the country are contributing to a transformation that is taking place in thousands of schools and classrooms. This revised edition builds on these exciting changes, offering ideas, strategies, and meaningful opportunities to bring science alive for students through a classroom or school science fair. It includes a new chapter on the science standards and lists relevant Web sites. Throughout the book, an icon directs readers to especially helpful Web sites listed with other resources in Chapter 15. Although the Web sites were current as of the writing of this book, with the ever-evolving nature of the Internet, some sites may change, others may be eliminated, and new ones will be added.

The untimely death of Isaac Asimov in 1992 created an enormous void in the scientific community. I have missed his counsel, support, and encouragement, but can assure readers that his unwavering philosophy and commitment to science education still permeate these pages.

The emphasis of this book is on the processes of learning. Here teachers and parents will discover an endless variety of creative learning possibilities. Students will find a cornucopia of mind-expanding, concept-building, and real-world experiences that will reshape their perceptions of what science is as well as what it can be.

Anthony D. Fredericks

The Point of Science Fairs

"The United States is, of all nations on Earth, the most technologically advanced."

But what do we mean when we say that? The phrase "the United States" is in many ways an abstraction. The United States is a region on the map, and it is also a region on the Earth's surface. It is a stretch of land with mountains and plains, rivers and deserts. It is a body of history and tradition, of laws and social custom. Yet all these things are empty of meaning if that is all that exists. So far, I have described only the background, the scenery of the play, the binding of the book. What we need, in addition, is the foreground, the actors of the play, the words in the book.

What the United States really is, more than anything else, is its population, the people that make it up, the people whose muscle and mind have formed, developed, and improved the nation over the generations and made out of what was once a wilderness, a mighty land that is the most technologically advanced on Earth.

But this means that it is the *American people* who are the most technologically advanced on Earth. Without our scientists, our engineers, our technologists, our construction workers, our skilled handlers of machinery, we could not maintain the technological superiority we possess. And that would mean we could not maintain our prosperity, our high standards of living, nor the strength we require to preserve our liberties and free way of life in a world that, for the most part, lacks all these things.

What must we do to preserve this technological advancement of our people? It would be foolish to concentrate on adult Americans, since they, for better or worse, have found their niches. They have received their education and chosen their work and social functions, and we must accept them as they are.

It is, rather, the children, who have not yet been educated, who have not yet chosen their work and social functions, on whom we must concentrate, for it is on them that we must rely to continue the technological advancement we need so much.

This is all the more so because the adult population is with us only temporarily. The decades pass, and our adults move into retirement. They are succeeded by the children, now grown up. Even before retirement, adults gradually find themselves out of touch with a technology that is rapidly improving and changing, while the children, growing up in that changed world, are at home in it.

It comes to this, then. If we are to keep the United States what it has been, and what it is now, we must concentrate on our children, for they are our greatest resource—our *only* resource in a way, for they will make all other resources possible.

And since it is our great technological expertise that keeps us comfortable and powerful, that means that our children must be well educated, well directed, well trained, in the direction that counts most—in the understanding of science and technology.

Of course, not everyone has the talent or the inclination to become a scientist or engineer, but those who do should surely receive the best education in that direction that is possible.

Again, we need to have millions of people who are skilled in other directions—farmers, entertainers, artists, writers, service workers—and yet even they should have some basic understanding of science and technology. After all, the important life-and-death decisions we as a nation must make concerning the ozone layer, acid rain, nuclear wastes, the greenhouse effect, pollution, all involve an understanding of science. Since the United States is a democracy, we must choose our own leaders and produce an enlightened public opinion that drives the leaders in appropriate directions. To know proper directions in which to encourage them to move, requires, these days, an understanding of science and technology, or democracy will prove a failure.

So if our technological advancement is to remain the best, the technological training of our youngsters must be the best, too. Unfortunately, there is general agreement that it isn't. We don't have enough teachers, especially teachers who understand science; and we don't have properly equipped schools. We must therefore strive to improve our educational procedures.

This can be done in many ways. Schools and teachers can use more money, more training, more equipment. In addition, however, there must be improvements in the very philosophy of teaching science.

It is insufficient merely to teach science as a block of facts, or as a set of rules. That can be unbearably dull for those who possess talent but lack a devouring fascination, for those who could do well if properly inspired, but *only* if properly inspired. We need something more than the words of a teacher presented to the student and then reproduced by the student on the test paper. That is so easily resented at the start and forgotten at the last.

It would be useful to involve as many students as possible in the actual *practice* of science. The important thing about science is not the advanced conclusions that can only be reached and understood by virtuosos. It is not the endless bodies of facts already observed and established. It is a *process,* a way of thinking and acting. What counts is the *scientific method* that makes all else possible.

This is what youngsters should learn—how to think scientifically, how to reason logically, how to make observations, how to gather and organize them, how to perform experiments and draw conclusions, how to make an intelligent guess in advance as to what those conclusions might be, and see them supported or rejected or left undetermined. And they must learn the joy and pleasure of doing all this for the sake of learning and not for awards—just as a game of football can be exciting even if you don't win.

This is the point of science fairs—to engage the interest of youngsters—to introduce them to the scientific method—to encourage them to understand science and possibly to become a scientist or an engineer—to help maintain the scientific lead and the prosperity of our nation—and, perhaps, to transfer that prosperity to the whole world.

And for this reason, Anthony Fredericks and I have prepared this guide to science fairs.

Isaac Asimov
(1920–1992)

A Note to Teachers

Science fairs have been part of the American education system since the early part of the twentieth century, offering students opportunities to display their scientific knowledge in engaging ways. Science fairs continue to offer students a showcase for scientific investigations and personal discoveries. They have been as much a part of the school curriculum as textbooks, standardized tests, and report cards.

The ultimate goal of a science fair is to encourage students to see how science works outside the classroom—how scientists investigate and learn about the world in which we live as well as worlds beyond ours. Today, more than ever, students need to understand and appreciate science: the scientific principles that influence their everyday lives as well as the scientific discoveries that add to their knowledge.

A science fair can be one of the most exciting parts of your classroom or school science curriculum. It can energize a science program and stimulate students' interest and participation. Unfortunately, a lack of resources for teachers and guidelines for students often result in lackluster participation and quickly designed projects.

If you tour a typical science fair you are likely to find:

- Too many volcanoes and solar system models; too little originality and planning

- A multitude of projects hastily constructed one or two nights before the science fair opened

- Projects that bear the unmistakable signature of Mom or Dad

- Poorly constructed projects that fall apart or collapse after a few days

- Projects that appeared last year, and the year before, and the year before that
- Frustrated parents and uninterested students
- A low level of participation by a single class or an entire school

This book is designed to be a convenient ready reference that helps teachers guide their students through the exciting and dynamic world of science. Its concept of effectiveness in a science fair is not based on the number of ribbons awarded and won. Rather, it places a premium on the processes of science—on helping to develop successful thinkers, not on trophies and awards. This book provides the ideas, strategies, and techniques you need to help students appreciate the world of science and their place in it— including a host of explorations and discoveries that will endure long after the last display has been taken down.

This book is a systematic guide to the design and development of a successful science fair. It is intended to stimulate higher levels of participation, well-designed and functional projects, an abundance of originality, and your students' deeper appreciation of how they can actively participate in the scientific world. In short, these ideas will stimulate:

- Greater student participation
- Greater creativity, originality, and overall quality
- Greater use of investigative skills and problem-solving activities
- A more positive attitude toward science

Everything you need to develop and promote a successful science fair is included. We suggest that you take time to read the different sections and to discuss them with your students. Let students know that taking part in the science fair is not only exciting in itself but will bring them a deeper appreciation of scientific processes and procedures.

As you discuss the different sections of this book with your students, invite their comments. Give them a chance to contribute their ideas to the science fair and you'll be guaranteeing their motivation throughout the entire event. Above all, make it clear to them that participation and involvement—not ribbons and trophies—are the crucial elements of the fair.

The ideas and strategies in this book are offered as suggestions, not as absolutes. We encourage you to make additions, subtractions, and changes according to the needs of *your* science fair. You may find it useful to duplicate these pages so that students and parents can use them as your science fair progresses from plan to reality. You, your students, and your students' parents will find valuable ideas garnered from science fairs throughout the country—procedures to ensure the success of any science fair.

The projects, strategies, and formats presented here are designed to give your students a fresh, exciting perspective on the scientific world. Their participation in a carefully crafted science fair can be their starting point for self-initiated investigations into the world around them. More important, this book and your science fair offer your students an enjoyable and worthwhile look at the wonders of science—a look that can last well beyond their projects and their school days.

A Note to Parents

Is your child about to participate in a science fair? Are you or your child wondering what to do or how to do it? Are you feeling nervous, anxious, or confused? Then *The Complete Science Fair Handbook* (Revised) is for you!

Science fairs can be one of the most exciting parts of the entire school curriculum. They provide children with an opportunity to explore the mysteries and marvels of the world in which we live and to develop an appreciation for the work of scientists. Yet for many families the announcement of an upcoming science fair can be an upsetting experience. Parents and children ask: How do we get started? Where do we get information? What should the final project look like? How much time do we need? Parents sometimes feel that schools do not provide them with enough information on how to plan and carry out a project. Without guidelines, families may be forced to improvise, hastily carting a last-minute project into the school gymnasium moments before the start of the science fair.

This book provides you and your child with a thorough, systematic approach to developing a successful science fair project—but it will not guarantee that your child will win a first-place ribbon or the grand prize trophy. The intent of this book is to help your child enjoy the *processes* of science discovery instead of focusing on winning a prize. Success is not defined in terms of how many awards your child's science fair project earns but by how well your child uses creative and investigative skills to discover more about his or her world.

The ideas, activities, and procedures outlined in this book will guide you and your child through an enjoyable and fascinating experience. Included are:

- How to plan and develop a project
- Sources of information and guidance
- Success factors that contribute to a display
- A host of potential projects and ideas
- How to assemble a project and write a report
- Standards used by science fair judges to evaluate projects
- A plan of action to ensure that the project will not have to be completed the night before the fair

Participating in a science fair can be a source of enormous satisfaction for your child, but it also means lots of planning and work. It is important that you offer encouragement and support whenever and wherever possible—not constructing the actual project but rather helping your child discover his or her own solutions. The value of a project lies in the amount of effort your child puts into it, not how much work you contribute. Learning to solve problems is something scientists do every day. That experience will be an important part of the work your child does in developing his or her project. Instead of building the entire project, we encourage you to build ideas in your child's mind. As your child works on a selected project, ask:

- What else could you contribute here?
- How would a scientist solve this?
- What other solutions can you think of?
- Is there another possibility?
- Can we look at this in another way?
- What else could we add?
- How do you feel about this project?
- Are you pleased with your effort? Why?
- Do we need anything additional?

We have designed this book to provide you and your child with many enjoyable experiences as you participate together in the selection, investigation, and construction of a science fair project. We encourage you to support your child's efforts at every step—guiding and encouraging whenever necessary. Your child's project can provide your entire family with many enjoyable hours together—time that will have a significant impact on your child's scholastic and personal growth.

TEACHERS: You may want to reproduce this and send it home with your students.

To Parents:
Introducing Our Science Fair

Dear Parents:

Soon your child will be taking part in an exciting school event—a science fair. Science offers children experiences in exploring beyond the classroom to understand more about their world. Investigating a selected science topic in detail can open up new vistas and a new appreciation for not only this planet but the worlds beyond.

 I would like to invite you to work along with your child as he or she selects, investigates, and reports on an appropriate area of science. With your interest and encouragement, your child can develop the skills and attitudes he or she needs to make this project a valuable experience. But do encourage your child to do most, if not all, of the work. Parents sometimes want to build an entire project, to make it "perfect." It is more important that your child wrestle with problems and try to solve them, because learning is in the doing. Guide your child whenever and wherever you can, but let the final project reflect your child's individual effort and design.

To help you in helping your child prepare for the science fair, I will be sending home instructions and suggestions throughout the coming weeks. These guidelines will give you and your child some ideas on how to create an effective project. Each sheet will show methods and processes that families can share throughout the science fair experience. Plan to take some time every now and then to talk these suggestions over with your child.

Remember that your child's success in our science fair will not be measured by ribbons, trophies, or certificates. Your child will succeed by learning and understanding more about science and how scientists work. Awards are secondary. The real goal of the science fair is stimulating your child's curiosity about the world.

I look forward to your participation in our upcoming science fair. Please call any time during the preparations and during the fair with your questions and suggestions. Let's work together so that our science fair will be a memorable and pleasant experience for your child.

Sincerely,

You can reach me at: (Phone) _____

between the hours of _____ and _____ (days) or _____ and _____ (evenings)

1 The National Science Education Standards and Science Fairs

In response to a growing concern about the state of science education in the United States, hundreds of teachers, scientists, science educators, and other experts worked together to outline the necessary ingredients for achieving scientific literacy.

This intensive examination resulted in the National Science Education Standards. The standards are based on the premise that science is an active process—that learning science is something that students do, not something that is done to them. Blending the ideas of "science as process" and "science as inquiry," the standards encourage a "hands-on, minds-on" approach. This approach helps students develop an understanding of science by combining scientific knowledge with reasoning and thinking skills.

Rather than a curriculum, the standards provide an outline for the development of science instruction. They are organized into six broad categories:

- Standards for science teaching
- Standards for professional development for teachers of science
- Standards for assessment in science
- Standards for science content
- Standards for science education programs
- Standards for science education systems

This section describes how science fairs support each of the standards in the *teaching* and *content* categories.

Teaching Standards

The six teaching standards describe what teachers should know and be able to do to help all of their students become scientifically literate.

Teachers of science plan an inquiry-based science program for students.

Science fair projects originate with questions drawn from students' interests and experiences. The content of science fair projects can be adapted to the needs and abilities of all students. Science fairs also encourage students to develop both short-term and long-term goals throughout the science program.

Teachers of science guide and facilitate learning.

Science fair projects encourage and model the skills of scientific inquiry. As students develop projects for a science fair, teachers challenge and support them while helping them to focus their self-initiated inquiries. Teachers also stimulate active discussions among students. During the development and assessment of a science fair project, teachers challenge students to accept responsibility for their own learning.

Teachers of science engage in ongoing assessment of their teaching and of student learning.

Science fairs provide teachers many opportunities to gather data about how well students understand science content. As students develop a science fair project, teachers also can guide them through a process of self-assessment. The information, observations, and interactions generated by science fairs can help improve the overall science program.

Teachers of science design and manage learning environments that provide students with the time, space, and resources needed for learning science.

Long-term science fair projects provide students with opportunities for extended investigations. As they develop a project for a science fair, students design their own learning environments. Science fairs encourage students to identify and use resources outside of school.

Teachers of science develop communities of science learners that reflect the intellectual rigor of scientific inquiry and the attitudes and social values conducive to science learning.

Science fairs help teachers celebrate the varied skills, ideas, and experiences of all students. A science fair project enables students to have a voice in decisions about their work and encourages them to take responsibility. Teachers have many opportunities to stimulate formal and informal discussions. Students learn the rules of scientific discourse when they are required to explain and justify their own science fair projects and assess those of other students.

Teachers of science actively participate in the ongoing planning and development of the school science program.

A successful science fair requires that teachers become involved in the planning and development of the school science program. Decisions about time and other resources dedicated to science education are natural consequences of the entire planning process. Science fairs also encourage teachers to work with each other across grade levels and subject areas.

Content Standards

The content standards outline what students should know and be able to do to be scientifically literate. They fall into eight broad categories.

Unifying concepts and processes

This standard addresses overarching concepts and processes, such as systems and change. Many science fair projects give students experiences that improve their understanding of these ideas.

Science as inquiry

This is what science fairs are all about. Science fairs stimulate the development of the abilities students need for scientific inquiry. Students learn to identify questions that can be answered by an investigation; design and conduct a scientific investigation; collect, analyze, and interpret data; and describe and explain their investigations. In addition, students are encouraged to think critically about evidence and to analyze alternative explanations.

Physical science, life science, earth and space science, and science and technology

These four content standards address different areas of science. Depending on what students choose to investigate, science fair projects can help them develop an understanding about the major principles in one or more of these disciplines.

Science in personal and social perspectives

Science fair projects can help students improve their understanding of personal health, changes in population, resources, changes in environments, local challenges, natural hazards, and science and technology in society.

History and nature of science

Science fairs provide teachers with many opportunities to promote science as a worthwhile human endeavor. Most science projects will give students insights into the nature of science, and specific projects can focus on the history of science.

Perhaps more than any other area of a science curriculum, science fairs hold the potential to address many elements of the National Science Education Standards. When science fairs are a standard part of the science curriculum, the value of science instruction is enhanced.

2 What Is a Science Fair Project?

A science fair project is a presentation of an experiment, a demonstration, a research effort, a collection of scientific items, or a display of scientific apparatus. It represents the efforts of a student's investigation into some area of interest and provides a way for the student to demonstrate the results of those investigations. A science fair project is a unique way for students to satisfy their own curiosity about the world around them and to pose questions for which they must seek out answers. It is a venture (and an adventure) into the world of scientific research that goes beyond lesson in the classroom or chapters in a book. Through the development of a science fair project, students gain a firsthand appreciation of the work of scientists and the value of their discoveries. Projects allow students to experiment, make decisions, form and re-form hypotheses, test and examine ideas, seek solutions, and most important, learn more about themselves and their world.

Science fair projects consist of three essential components: a display unit, exhibit materials, and a written report.

Display Unit

The display unit forms the background for the project. It should be built of sturdy materials to provide a structure for a vertical display of graphs, charts, photographs, and other printed information. Usually three-sided, it includes the name of the project as well as other information that is vital to observers.

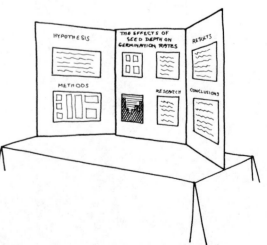

Exhibit Materials

The exhibit materials consist of items collected or demonstrated by the student, a set of apparatus, or the experiment the student carried out during his or her investigation. Display materials give the science project a three-dimensional effect and allow others to observe the actual materials involved in the student's investigation.

Written Report

It is important for students to keep a written record of their investigations. This record outlines the original problem the student chose and the means and methods used to investigate it. The written report should be accurate and easy to read, and it should give a clear summary of the entire project.

The final project a student sets up for a science fair is limited only by his or her imagination and curiosity. Projects have taken many forms and designs over the years—each the end product of a question a student had about the world of science. Helping students develop and expand a deeper appreciation for scientific investigations may be the most important by-product of a science fair project.

3 Keys to a Successful Project

Success in a science fair can be judged by a number of standards, but it should *not* be measured by ribbons, trophies, or other awards. If you, the student, have selected a topic, investigated it according to your own design, and reported the results of that investigation in the form of a display and a written report, you have succeeded. Winning "first place" or being "grand champion" is certainly praiseworthy, but your goal in taking part in the fair should be to investigate an area of interest and to discover new things about the area you choose. The satisfaction of making these discoveries will last far longer than blue ribbons or gold medals.

Here is a list of factors that you can use to evaluate your project and that your parents and teachers can use in judging it. Use the list throughout your investigations and also upon completion of the project to gauge the appropriateness of project features. Put a check mark beside each fact to assess the completed project.

❏ **Does the project represent the student's own work?**
Although students may receive help in investigating their topic and designing their respective projects, the final effort must be the student's—not that of a scientist, teacher, parent, or other adult.

❏ **Is the project the result of careful planning?**
Successful projects cannot be accomplished overnight. They are the outcome of a systematic plan of action carried out over a period of time. A hastily constructed project undermines the value of the science fair.

❏ **Does the project demonstrate the student's creativity and resourcefulness?**
Students should be permitted and encouraged to contribute their own ideas and ingenuity to the design and development of a particular project.

❑ **Does the project indicate a thorough understanding of the chosen topic?**
Students need to investigate their chosen area as completely as possible. Doing so will take time. The project must reflect the results of those investigations done over an extended period.

❑ **Does the project include a notebook, written record, or final report?**
The display should include a written summary of the investigation. Such a record provides observers with additional information on the subject and documents the student's work.

❑ **Does the project include a number of visual aids?**
Photographs, charts, diagrams, graphs, tables, drawings, or even paintings liven up any display and make it more interesting.

❑ **Is the project sturdy and well constructed?**
Using the proper materials and being careful in assembling a project are important, particularly if the display will be standing for several days. It must be within the required size limitations and should reflect a degree of permanence.

❑ **Are all signs and lettering neat and accurate?**
The quality of a display is often judged by the attractiveness of signs, titles, and written descriptions.

❑ **Does the project meet all safety requirements?**
When electrical items, specimens, or chemicals are used in a display, care must be taken to ensure the safety of any observers. The display of any live organisms is discouraged.

❑ **Is the display three-dimensional?**
In addition to the background and accompanying written report, the inclusion of samples, apparatuses, collections, or other items is vital to the project. These should be attractively arranged in front of the background display.

❑ **Is all information accurate?**
Any data gathered from outside resources, such as printed
materials or interviews with experts, and data obtained
from experiments, must be presented accurately. All
questions about data must be resolved before including
them in a report or on a display.

❑ **Does the display present a complete story?**
The student should illustrate the topic chosen for
investigation, what was done during the investigation, the
results, and a conclusion. In other words, the project
should have a beginning, a middle, and an end.

4 For Teachers: How to Make Your Science Fair Successful

Putting on a science fair can intimidate even the most well-intentioned educator. The amount of time and work that precedes a successful fair often seems enormous. Nevertheless, the payoffs can be tremendous: students who gain an increased awareness of the importance of science in their lives and are able to investigate areas of interest that add to that knowledge base.

Here are some suggestions you may wish to consider as you prepare students for an upcoming fair. Use these ideas throughout the weeks preceding and during the fair to create a memorable event that students will look forward to in succeeding years.

The reward is in the doing.

Emphasize to students that the object of the science fair is not to win first place or a blue ribbon but rather to participate. Some students may suspect that you don't mean it. You can ignite their interest by announcing that everyone who enters will receive some sort of recognition—whether it will be in the form of a letter of appreciation or an announcement in the school newspaper. In short, everyone who enters wins.

Tie the fair to other subjects.

Often science fairs are done in isolation from other areas of learning. Try to incorporate some of the strategies in this book throughout the school day. For example, writing the project report can be part of language arts lessons. Research for the project can be a part of reading lessons. Math skills can be reinforced through the measurement or estimation of project amounts and quantities. Social studies lessons can include an examination of famous scientists and their contributions. In short, a "whole curriculum" approach to science fairs can maximize interest and participation.

Involve the whole school in the team.

Work along with other individuals in your school to provide a team approach to the science fair. For example, ask the librarian to prepare a special display of books about science experiments or famous scientists. Have the principal visit class to talk about other science fairs. Encourage the reading specialist to present a story or lesson with a science theme. Ask a colleague to visit your room to share a science-related hobby or area of interest.

Involve the community.

Get students involved in promoting the science fair. Have them:

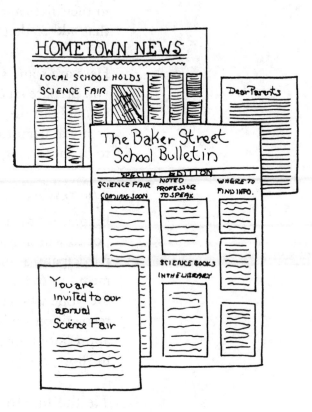

- Create a classroom or school newsletter.

- Distribute letters and notes to parents.

- Send invitations to the principal or other school officials.

- Write news releases about the fair and send them to the local newspaper.

Students may wish to extend an invitation to a local scientist or college professor to visit your class and explain a concept or demonstrate a scientific principle. Invite parents and other community people to bring in rock collections, telescopes, or vacation slides to share with your students.

Students may wish to promote the science fair throughout the school by setting up a panel discussion in different classrooms (selected students can be designated experts in a particular field). In addition, students may want to prepare a demonstration for an all-school assembly, conduct a question-and-answer quiz show via the intercom system, or produce a special videotape for younger children to view.

Finally, special displays or projects set up around the school, especially if they are constructed by your students, can go a long way toward stimulating interest and high levels of participation.

Keep parents informed.

This book provides you with a number of forms, letters, and notices you can send to parents throughout the weeks prior to the science fair. Let parents know that you are eager to provide any additional information they need. You may wish to set up a special newsletter for parents, call parents periodically to offer assistance or guidance in project preparation, or meet with them after school or in a special evening meeting. Above all, try to alleviate their fears and uncertainties about the upcoming fair. In addition, impress upon them the necessity of supporting their child's efforts, not constructing their child's project.

Establish a long-range timetable and stick to it.

The major factor in unsuccessful science fair projects is lack of proper planning. Use the timetables in this book to ensure that students are allotting enough time for sufficient investigation of their areas of interest. Make sure your students—and their parents, too—understand that science projects must be investigated and constructed over a period of time and cannot be done in the one or two evenings preceding the fair.

Don't let "fear of science" stop you.

Not being scientifically oriented shouldn't discourage you from conducting a science fair. Join forces with another teacher: Establish a partnership and plan the event jointly. Locate parents and community members with science backgrounds. Invite them to visit the classroom regularly to assist you and the students in designing individual projects as well as the entire science fair. There is no need to work in isolation—a joint effort may be just the ticket for a productive and successful event.

Keep your principal or supervisor informed.

Good communication is essential to a successful science fair. Get your administrators involved in the dynamics of the fair and frequently solicit their advice on how to promote your students' efforts.

Keep it exciting; make it fun!

Above all, demonstrate by your own attitude that science fair projects are fun. Your attitude toward the fair goes a long way toward ensuring its success. Let students know how enthusiastic you are about the event and they will match your enthusiasm.

More Tips for Success

Teachers from around the country have suggested the following tips for a successful science fair—whether a classroom or district-wide event. Consider these ideas in the preparation and planning of your own science fair.

- Plan ahead. Establish the date of the fair early. Make sure it does not conflict with other events on the school calendar.

- Establish a planning committee that includes students. Assign roles and responsibilities to individuals and/or subcommittees. One person cannot do all the work associated with a science fair.

- Seek permission from school or district administrators.

- Secure an appropriate exhibit area (e.g., cafeteria, gym, community room) well in advance.

- Determine how many monitors and assistants you will need during the week and day of the fair and schedule the times they should be there. Also obtain the assistance of the school's custodial staff.

- If projects are to be judged, make early arrangements for an adequate number of judges. Provide them with dates, judging criteria, directions, times, and other information. Call to confirm their participation several days before the fair.

- Print and distribute forms early in the process. These include parent permission slips, judging criteria, display requirements, student registration cards, student planning guides, and others.

- Arrange for publicity early to keep people in and out of the school up-to-date on the progress of the fair.

- Plan a visitation schedule for the fair that specifies when other classes can visit and when it is open to parents and people in the community.

- Get parents involved early. They can serve on subcommittees, assist as fair monitors, help with publicity, design name tags and project labels, or volunteer in other ways on the day of the fair.

- Work very closely with students throughout the entire science fair process. Provide extended periods of time within the regular classroom curriculum for students to work on their projects.

- Consider conducting mock interviews and mock evaluations of projects to help students become comfortable with these elements of a fair.

- The day before the fair opens, double-check everything—from the delivery times for projects to the scheduling of visits to the availability of electrical outlets near displays that require them. Conduct a "walk-through" of the exhibit area.

- After the fair, hold a debriefing session with committee members to evaluate all components of the fair. What can be done to make next year's fair even better? Be specific and detailed in the evaluation—it will pay off in the success of next year's science fair.

5 Two Science Fair Timetables

Successful science fair projects take planning. Trying to put a project and report together a few nights before the scheduled opening can lead to disaster. Besides submitting a hastily constructed project, the student fails to develop an appreciation for the time and effort scientists need to conduct their investigations. Planning a project well in advance allows sufficient time for the necessary research, the construction of the display, the writing of the report, and the assembling of the final project. It also provides some leeway should difficulties arise in research or in obtaining vital materials.

The timetables that follow can be used by science fair coordinators, teachers, parents, or students. A 12-week and a 6-week timetable are provided, but students should use the 12-week if at all possible. Twelve weeks provide sufficient time to design a project, gather necessary data, develop a written report, and follow through on all components of a successful science fair project. Of course, circumstances may be such that the 6-week schedule must be followed.

As soon as you know the date(s) of the science fair, use a calendar to count back 6 or 12 weeks from the opening date to find when preparations should begin. For example, if a science fair is scheduled for April 1, a 12-week preparation should begin on January 1.

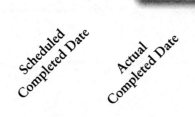

12 -Week Timetable

Date of the science fair _____

Date to begin working on the project (count back
12 weeks from the science fair opening date) _____

_____ _____

Week 1

- Choose a topic or problem to investigate.
- Make a list of resources (school library, community library, places to write, people to interview).

_____ _____

Week 2

- Select your reading material.
- Begin preliminary investigations (select Web sites).
- Write for additional information from business firms, government agencies, and so on (see Resources).
- Start a notebook for keeping records.
- Write down or sketch preliminary designs for your display.

_____ _____

Week 3

- Complete initial research.
- Interview experts for more information.
- Decide how to set up your investigation or experiment.
- Decide what materials you will use in the display. Make a list.
- Set up experimental design.

_____ _____

Week 4

- Begin organizing and reading the materials sent in response to your letters.
- Decide whether you need additional material from outside sources (i.e. Web sites).
- Begin collecting or buying materials for your display.
- Begin setting up your experiment or demonstration.
- Add information to the project notebook as you get it.
- Start your collection or experiment.

_____ _____

Week 5

- Learn how to use any apparatus you need.
- Continue recording notes and observations in your notebook.
- Set up outline for written report.

Week 6

- Gather preliminary information in notebook.
- Work on first draft of written report.

Week 7

- Start assembling display unit.
- Continue recording notes.
- Check books, pamphlets, magazines, and Web sites for additional ideas.
- Verify information with experts (teachers, professors, scientists, parents).

Week 8

- Begin designing charts, graphs, or other visual aids for display.
- Take any photographs you need.
- Record any observations on experiment.
- Begin preparing signs, titles, and labels for display unit.

Week 9

- Have photographs developed and enlarged.
- Talk with experts again to make sure your work is accurate and on schedule.
- Begin writing second draft of your report.
- Continue recording observations in notebook.

Week 10

- Write text for background of display and plan its layout.
- Complete graphs, charts, and visual aids.
- Finish constructing your display.
- Work on final draft of written report.

Week 11

- Complete your experiment or collection.
- Write and type final copy of written report.
- Do lettering of explanations and mount them on your display.
- Mount graphs, charts, drawings, and photographs.
- Assemble apparatus or collection items; check against your list.

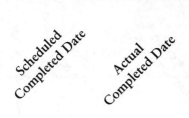

_____ _____

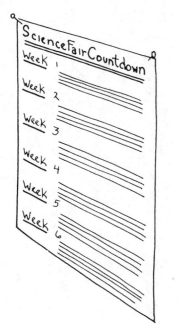

Week 12

- Proofread written report.
- Set up display at home and check for any flaws (leave standing for 2 days).
- Carefully take display apart and transport it to science fair site
- Set up display.
- Check and double-check everything.
- Congratulate yourself!

6-Week Timetable

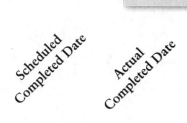

Date of the science fair _____

Date to begin working on the project (count back 6 weeks from the science fair opening date) _____

_____ _____

Week 1

- Choose a topic or problem to investigate.
- Check resources in school or community library.
- Contact experts in the field (i.e. Web sites).
- Gather all written material you can find on the topic.

Week 2

- Begin putting your project notebook together.
- Start collections or experiment.
- Begin designing display unit.

Week 3

- Begin building display unit.
- Design all visual aids.
- Take the photographs you need.
- Complete your research (check Web sites).
- Consult with experts (teachers, professors, scientists, parents) to check your progress.
- Write first draft of report.

Week 4

- Continue collecting items for display.
- Continue your experiments.
- Set up your apparatus and test it.

Week 5

- Write second draft of report.
- Construct background for display.
- Design and assemble graphs or charts.
- Complete lettering for display unit and mount it.
- Double-check your written data.
- Complete experiment and record data.

Week 6

- Write and type final report.
- Set up display unit at home and test.
- Transport display to science fair site, set it up, and test it.

6 Helping Students Select a Topic

Choosing an appropriate topic for a science fair project is often the most difficult part of the entire process: So many topics to choose among, a wealth of information available. No wonder students are bewildered. Teachers and parents can help provide direction and narrow down choices, but the final choice of topic should be the student's. His or her motivation will be a critical factor in the successful completion of the project. Parents and teachers may pose these questions to help students refine their choices and decide on the most suitable area to explore.

Interests

- What kinds of things do you enjoy doing?
- What area of science interests you the most?
- If you could be a scientist, what would you like to do?
- What are your hobbies or free time activities?
- What do you like to do on rainy days?
- What kinds of books do you like to read?
- Which movies or TV shows might give you ideas or information?
- What are your special skills or talents?

Difficulty Level

- How hard will this topic be for you to understand?
- What problems have you had with this subject before?
- Are you familiar with this topic or is it brand new?
- Do you think you will need to gather a lot of outside information?
- Will you be able to work in this area for 12 (or 6) weeks and still be interested?
- What special tools or apparatus do you think you'll need?

Time

- Will you be able to spend some time on this project every week for 12 (or 6) weeks?
- How long do you think you will need to gather information on this topic?
- Are you interested enough in this subject to spend a great deal of time on it?

- Will you need to set up a special schedule to complete all the things you need to do?
- Do you have enough free time at home to work on the project?

Materials

- What special materials do you think you'll need for this project?
- Do you have those materials at home or will you need to buy them?
- Will you need to construct anything complicated?
- Will you need help in putting the display together?
- Will you need to order any materials through the mail?
- Will you be able to buy materials in local stores?
- Will your materials be inexpensive or costly?

Guidance

- How much help will you need with your project?
- Will you be able to do most or all of the work yourself?
- Will you need to consult any experts in your chosen field?
- How much involvement will your parents have?
- Will you be able to build the display unit on your own?

Safety

- Will you be able to follow all safety rules in putting your project together?
- Are there any dangers from equipment or materials associated with your project?
- Will there be any dangers to observers of your project?
- Will there be any danger to you at any time during the investigation of this project?

Oftentimes, students select a topic simply because everyone else has selected it (that's why there are so many volcanoes and solar system displays at most science fairs). Students need to understand that the choice of an appropriate topic depends on several factors that must be discussed and agreed on before the project is begun. Of course, the primary criterion will be: Is it something the student is truly interested in pursuing? Allowing students to explore self-chosen areas of interest will be a major factor in making the science fair a positive learning experience.

When students have trouble choosing a topic or project for a science fair, an "idea generator" can help. When students generate their own ideas—when they have that sense of ownership—they become more actively engaged in their science fair project.

Two powerful tools for generating ideas are discussed here: "What if" questions and "I wonder" statements.

What If...?

Research suggests that students at all grades should be exposed to more questions that generate creative thinking. By doing so, teachers and parents can help students approach any scientific task—including selecting a science fair topic—with curiosity and enthusiasm.

One of the best ways to help students identify a science fair topic is through the use of "what if" questions. By tagging those two simple words to an idea, teachers can turn topics, issues, or principles from the science curriculum into creative questions. For example, the question "What if all animals were red?" might spur a student to investigate a variety of possible topics: animal camouflage, predator/prey relationships, adaptation of specific species, evolution, environmental changes, or coloration. "What if" questions introduce many possibilities to explore.

It is important to remember that there are no right or wrong answers to "what if" questions. They get students to look at science with a creative eye, to think about possibilities rather than absolutes.

The following list of "what if" questions can be used to help the class or individual students generate a topic for a science fair project. Consider the list a starter kit, with the idea that teachers, students, and parents may add questions of their own.

Plants

- What if plants were not watered?
- What if plants had no seeds?
- What if plants responded to music?
- What if plants received no sunlight?
- What if plants grew in human hair?
- What if plants grew upside down?
- What if we had no plants to eat?
- What if all plants lived in the desert?
- What if all flowers were the same color?

Animals

- What if all animals were the same color?
- What if all animals were the same size?
- What if there were more mammals than insects?
- What if the largest animal were an insect?
- What if someone found a dinosaur fossil in our town?
- What if animals never lived in herds or groups?
- What if all animals ate the same food?
- What if humans could swim like fish?
- What if all animals lived underground?

Environmental Studies

- What if there were no pollution?
- What if factories were outlawed?
- What if the air were always dirty?
- What if people were banned from driving their cars one day a week?
- What if smoking were illegal?
- What if an efficient method of trash disposal were invented?
- What if everybody recycled?

Matter

- What if everything in our house were the same weight?
- What if everything in our house were the same color?
- What if we could change something from a liquid to a solid in a new way?
- What if there were no atoms?
- What if matter could not be changed chemically?
- What if there were five states of matter?
- What if mass and weight were the same thing?

Energy

- What if light could travel twice as fast as it does?
- What if there were no light?
- What if there were only five simple machines?
- What if there were only one form of heat?
- What if humans had to produce all energy?
- What if static electricity did not exist?
- What if we could not use electricity one day a week?
- What if sound could be heard on the surface of the moon?
- What if sound energy moved in a straight line rather than in waves?
- What if we didn't have magnets?

Earth

- What if there were no erosion?
- What if there were no weathering?
- What if all the world's oceans were the same depth?
- What if there were a volcano in our town?
- What if earthquakes occurred where we live?
- What if there were only two types of rocks?
- What if there were no mountains?
- What if all of the water in the world were polluted?
- What if all nonrenewable resources were suddenly depleted?

Space

- What if there were another solar system?
- What if we had no sun?
- What if scientists discovered another planet in our solar system?
- What if we could travel to Mars?
- What if an asteroid hit the earth?

- What if our planet stopped revolving around the sun?
- What if all the planets were the same distance from the sun?
- What if we could travel to any planet?

Weather

- What if it always rained?
- What if it never rained?
- What if a hurricane or tornado hit our town?
- What if there were no seasons?
- What if we lived in the Southern Hemisphere?
- What if there were no winds?
- What if the sun heated the earth equally in all areas?
- What if we had the same temperature night and day?
- What if the skies were always cloudy?

Human Body

- What if our digestive tract were twice as long as it is?
- What if we had a faster heartbeat?
- What if we never ate?
- What if we could survive on one breath per minute?
- What if we had no teeth?
- What if we had more than two arms, two ears, two legs, two eyes, and two lungs?
- What if we ate only sugary foods?
- What if we ate only carbohydrates?
- What if we had no brain?
- What if we could choose our body shape and size?
- What if only selected individuals could reproduce?
- What if we did not have a skeleton?

I Wonder....

"I wonder" statements are another technique to help students choose an area of investigation for a science fair project. This strategy encourages students to think like scientists. Scientists often wonder why something happens the way it does. That sense of wonder often leads to a hypothesis and an investigation to test it.

A teacher might begin the day or a science lesson by asking each student to create a sentence beginning with the words "I wonder...." It is important to take time to focus on the topics, interests, or concerns that students raise and to share the possibilities with the entire class. These statements also can become the impetus for science lessons throughout the year.

There are other ways to generate ideas with "I wonder" statements. A teacher could invite students to record an "I wonder" statement in a personal journal. This can be done on a regular basis, such as every morning or at the end of each week. Students can post their "I wonder" statements on a bulletin board or share them with classmates in group discussion sessions. As students listen to or read the "I wonder" statements of others over an extended period, they will see a wide variety of topics available to explore in a science fair project.

To encourage students to create their own "I wonder" statements, teachers should model the technique throughout the day. Teachers can model several "I wonder" statements and then invite students to create their own. Again, there are no "right" or "wrong" statements; students should be encouraged to consider any number of possibilities.

"I wonder" statements create an enormous bank of scientific ideas from which students can generate creative science fair investigations.

Teachers and students should add their own "I wonder" statements to this brief list of suggestions.

- I wonder how plants grow.
- I wonder how plants reproduce.
- I wonder why plants need sunlight.
- I wonder why some plants aren't green.
- I wonder why dinosaurs died out.
- I wonder why there are so many ants in the world.
- I wonder why some animals are endangered.
- I wonder why some animals don't have backbones.
- I wonder why there is pollution.
- I wonder if people really want to recycle.
- I wonder if we can clean up our toxic wastes.
- I wonder if smoking is harmful.
- I wonder how water freezes.
- I wonder how scientists know about atoms.
- I wonder how many different gases there are in the world.

- I wonder how I can measure a liquid.
- I wonder why magnets work as they do.

- I wonder how electricity moves through wires.
- I wonder how sound is produced.
- I wonder how a wheel and axle work.
- I wonder why there are so many active volcanoes.
- I wonder how mountains are formed.
- I wonder about how scientists explore the world's oceans.
- I wonder where diamonds are found.
- I wonder how scientists measure distances in space.
- I wonder what the nearest star is.
- I wonder what will happen to the sun.
- I wonder how fast a meteor can travel.
- I wonder how hurricanes are created.
- I wonder why it rains so much in some places and not in others.
- I wonder how scientists use weather instruments.
- I wonder how clouds are formed.
- I wonder how tall I'll grow.
- I wonder what kinds of foods I should be eating.
- I wonder what happens when I break a bone.
- I wonder how much exercise I should get each day.

8   Suggestions for Projects

As mentioned already, many students find that selecting a topic is one of the most difficult parts of the entire science fair experience. With so much to choose from, students are often overwhelmed by the enormous variety of potential subjects they can explore or examine. Making a decision on the most appropriate topic can be a time-consuming and mind-boggling experience, not just for students, but for teachers and parents, too.

The lists that follow are intended to provide a selection of potential topics for students to investigate. These lists are not intended to include all possible topics but rather to offer a diversity of subjects from which students can choose. It is important to note that the topics have been arranged by grade levels to make selection easier. To do this, we consulted leading science texts at each grade level to identify topics included in the science curriculum for that grade. That is, each of the topics on a particular list is normally taught at the grade for which it is listed. In addition, several experts in different fields were consulted for their ideas on potential topics for each grade level. Thus, students have the opportunity to select a topic commensurate with their interest as well as their specific grade level.

This does not mean that students should be restricted to any single list. Encourage students to select a topic in keeping with their desires and interests, not necessarily because it appears on a list headed by their particular grade. As teachers and parents guide students in the selection process, encouraging the exploration of a topic not commonly taught at the student's grade is certainly appropriate. These lists represent many grade levels and many ability levels, within grades. We've provided something for everyone across the grades as well as within a specific grade so that gifted, on-level, and remedial students will find a host of scientific possibilities to investigate.

These lists also provide a diversity of possibilities in each of the three major areas of science: life science, earth and space science, and physical science. Thus, students have an enormous range of topics in any discipline. Teachers and parents should assist students in examining the widest possible range of topics before making their final choice. Just by exploring possible topics, students discover the magnificence and diversity of the world of science and the enormity of its possibilities.

The potential topics listed within each grade level are organized into five categories: experiments, demonstrations, research, collections, and apparatus. These categories are not absolute, they overlap; but focusing on the types of projects possible helps students narrow down their selections.

For more suggestions for science fair topics, refer to Web Resources in Chapter 15. Look for the section titled Science Fair Project Ideas.

Experiments

The type of project most often presented at science fairs is the experiment. These presentations allow students to pose a problem, design an experiment to investigate that problem, record and report their results, and make conclusions based upon those results (see the section on scientific method in Chapter 10). The final project is a display of steps the student took, any successes or failures, and the implications of the data.

Demonstrations

In this type of project students demonstrate a particular scientific principal or fact. The demonstration should be self-contained; that is, observers can operate or manipulate any controls, switches, or devices needed for the demonstration. Students may wish to demonstrate how something works, a science phenomenon, or how something is created naturally or in the lab.

Research

In a research project, the student investigates a chosen area of science by consulting *primary* sources. That is, students will need to consult reading materials from libraries, museums, government agencies, and the Internet. In addition, they should interview experts: scientists, health care workers, county agents, shop forepersons, and so on. Encourage on-site investigations at labs, factories, a printing plant, a farm, or a fish hatchery. The intent is for the student to explore a scientific area in depth and detail and to report findings in a vivid, interesting way through the project.

Collections

Collections are an assembly of items such as seashells, birds' nests, or telephone parts that show variety and diversity within a chosen area of science. Usually, collection projects will result from a hobby or other free-time activity. Collections need to include as many samples as possible to represent the magnitude of the topic.

Apparatus

In this type of project students display some kind of scientific apparatus or instruments and describe their use or function in detail. The project should enumerate the importance of the apparatus for both scientists and the general public. Descriptions of how each apparatus is used within or outside the scientific community would also be appropriate.

Grades 4 to 5

Experiments

- Test any responses to real and artificial sweeteners.
- Effects of temperature changes on fish.
- Do preservatives stop bread mold from growing?
- How leaves lose water.
- Effects of sunlight on plants.
- Effects of crowding on plants.
- How changing the fulcrum affects a level.
- What fabrics make good insulators?
- How do charged objects act toward each other?
- Materials that are the best conductors of electricity.
- Effects of height of a swinging mass on its energy.
- How are crystals formed?
- Removing salt from water.
- Which foods contain starch?
- Which sense organ can detect the greatest variety of sensory information?

Demonstrations

- Construct a clay model (with cutaway sections) showing the three layers of the earth.
- Create your own fossils, using plaster casts.
- Make a model of the ocean's floor, labeling each part.
- Construct a model of the eye showing its different parts.
- Where different flavors are tested on the tongue.
- Using modeling clay, make a cross section of the skin.
- What does a magnetic field look like?
- Using a graduate, measure the volume of several objects.
- Set up a box with two holes in it (for hands to reach in) containing unknown objects. Participants reach inside and try to guess what the objects are by feeling them and describing their characteristics.

- Testing minerals for their various properties.

Research

- Show how living things depend on one another through food chains.
- Use food webs to show how members in a community get their energy.
- Illustrate how animals live underground.
- What are the types of jobs bees have in a honeybee colony?
- How are bees helpful to humans?
- Ants and their jobs.
- Show examples of parasite and host relationships.
- Diagram the parts of trees or flowers.
- The life cycle of nonseed plants.
- Prepare a nature guide to plants and trees on the school grounds or in your neighborhood.
- How plants make food.
- How animals and plants adapt in order to survive.
- Types of bird beaks and their function.
- Why animals hibernate.
- Pick a career in science and tell about it.
- Examples of potential and kinetic energy.
- Learn about insulators and conductors.
- How rocks are formed.

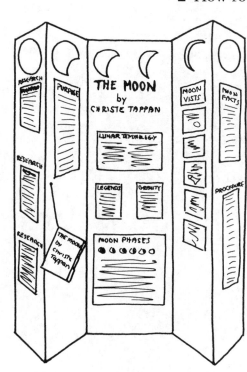

- Uses of rocks and minerals in everyday life.
- The formation of coal.
- Chart the Gulf Stream or any other major warm- or cold-water current.
- Using resources from the sea: advantages and disadvantages.
- All about the wind chill factor.
- How air temperature changes.
- The Beaufort wind scale.
- Chart similarities and differences between the planets (temperature, distance from the sun, moons, length of day, and so on).
- Everything you wanted to know about Saturn (or Mars, or Neptune, or Uranus...).
- Record local temperatures (at regular intervals) throughout the day for several weeks.

- Compare predicted weather with actual weather.
- The digestive system and how it works.
- From cells to systems.
- The human eye and how it works.
- What is color blindness?
- The history of measurement.

Collections

- Clay models of animals that live in groups.
- Start your own ant colony.
- Collections of any of the following: leaves (fall foliage or green), pinecones and/or needles, weed seeds, plants that reproduce without seeds.
- Make casts of animal tracks.
- Birds' nests: collect the materials used in building nests.
- Monocot and dicot seeds and/or flowers.
- Simple machines used in everyday life.
- Start a collection of rocks found in the area.
- Collect some common minerals.
- Use pictures to show examples of animal populations: herds, colonies, schools, and so on.
- Display pictures of herbivores, carnivores, and omnivores.
- Collect items that show different forms of energy (chemical, light, sound, heat, electrical, mechanical).
- Demonstrate different types of animal teeth.

Apparatus

- Construct a homemade thermometer.
- Series and parallel circuits.
- How to make electromagnets.
- Make each of the following and describe how it works: barometer, anemometer, wind vane, rain gauge.
- Construct a balance and invent your own measuring system to measure matter.
- Make an electrical question board.

Grades 5 to 6

Experiments

- How water rises in different kinds of plant stems.
- Does a temperature change in water affect a fish's rate of breathing?

- What are some behaviors of earthworms?
- What kinds of foods do certain types of birds prefer?
- Is air matter?
- Forming compounds.
- How heating water affects the rate at which materials dissolve.
- Factors affecting how fast liquids will mix.
- Boiling points of liquid substances.
- Effects of particle size on how fast a solute dissolves.
- How to make water wetter.
- What metals and/or materials will rust?
- Good and poor conductors.
- Effects of circuit type on the brightness of lightbulbs.
- Effects of object color on how warm it gets.
- Where is the best place to position solar heating units?
- Can the wind be used to make electricity in the area where we live?
- Is a solar collector a feasible way to heat water?
- Are there solid particles in the air we breathe?
- Effects of dilution on reducing water pollution.
- What materials are biodegradable?
- Effects of item color on the amount of solar energy it absorbs.
- Observe cloud patterns for several weeks and try to predict the weather. How accurate are your predictions?
- Controlling eye blinking.
- Effects of removing minerals from bones.
- Effects of different kinds of physical activity on pulse rates.
- Factors effecting condensation.

Demonstrations

- Plant a dozen bean seeds. After they have sprouted, describe the changes that have occurred at intervals of five days.
- Make a model of a cross section of a leaf.
- Using real flowers, observe with a magnifying glass and locate their parts. Make a flower and its parts from modeling clay.
- Sprout seeds without using soil.

- Build an earthworm farm.
- Create and label the parts of an imaginary insect. Include all the body parts needed by a real insect.
- Where does water come from?
- Illustrate and name the birds that use discarded birds' nests.
- Collect materials that birds use and make your own set of birds' nests.
- Pick an animal community and display it in a diorama.
- Create a terrarium.
- Create a model of an atom.
- Create models of a variety of common molecules.
- Have participants guess the contents of a wrapped box by using indirect evidence.
- Construct the two types of circuits.
- Compare and contrast the different types of batteries.
- How glaciers change the land.
- Construct a spider web (using twine or rope). Attach models of different spiders or important facts about arachnids.
- Make papier-maché globes and sun. Use them to demonstrate the changing seasons.
- Construct a relief map of North America. Name and label the major air masses.
- Use cotton balls to make cloud formations.
- Cover boxes with black paper and punch holes through the back in patterns that represent constellations. Put a light source behind the boxes.
- Construct models of constellations using modeling clay, papier-maché, or other appropriate materials.
- Make a working model of muscles and bones in the arm or leg.
- Prepare cutout drawings of the different major parts of the skeletal system. Challenge participants to assemble them.
- Using a life-sized paper model of a body and two different colors of yarn, show the circulatory system.
- Set up a display to test the lung capacity of different individuals.
- Use an art medium of your choice to make a model of the human heart.
- Demonstrate eclipses of the sun and moon.
- Using models, show the causes and effects of tides.

Research

- Study a local bird population, recording the number and types of birds that visit a feeder. Graph your findings.
- How plants get materials for food making.
    - What is photosynthesis?
    - What is respiration?
    - Compare photosynthesis and respiration.
    - How flowers produce seeds.
    - Who are the invertebrates?
    - The worm family.
    - All about echinoderms.
    - The structure of a fish.
    - Chart the similarities and differences between reptiles and amphibians.
    - The life cycle of a frog.
    - The snakes of our area.
    - Interesting facts about birds.
    - The structure of bird bones and feathers.
    - How milk is pasteurized.
    - What are mammals?
    - Everything you wanted to know about the platypus.
    - Gestation periods of different mammals.
    - All about koalas (or other Australian animals).
- Characteristics of the five groups of vertebrates.
- A whale of an assignment all about whales.
- Endangered and threatened species of animals and/or plants.
    - Environmental effects on the size of animal populations.
    - Succession in an ecosystem.
    - Research a science career that interests you.
    - Make a time line of a famous scientist's life.
    - So you want to be a zoologist (or botanist, microbiologist, chemist ...)?
    - Graph the number of kinds of animals in the major animal groups.
    - Launch helium-filled balloons (including your name and address on a return postcard). Record and chart information received from respondents on distance traveled, wind direction, and elapsed time.

- Common uses for elements.
- How the symbols for elements evolved.
- Solutions and suspensions found around me.
- Products that result from chemical changes.
- Electricity—the energy around us.
- Uses of parallel and series circuits.
- The impact of Michael Faraday's work.
- Rules for electrical safety.
- Sources of energy.
- Major locations of coal, oil, and natural gas.
- How are the major sources of energy used to produce electricity?
- The greenhouse effect: How will it affect the Earth?
- Nuclear energy: benefits and problems.
- Ways to conserve energy.
- Physical and chemical weathering.
- How to prevent soil erosion.
- Water-treatment plants.
- Effects of acid rain.
- Major sources of air, water, and land pollution.
- Preventing pollution: different ways, different methods.
- All about air masses.
- The global wind belts.
- Severe weather phenomena.
- Cold and warm fronts: What are they?
- Weather records and extremes.
- What is astronomy?
- The life cycle of a star.
- The skeletal system.
- Joints of our bodies.
- Activities of voluntary and involuntary muscles.
- The artificial heart.
- The function of the respiratory system.

Collections

- Observe earthworms by starting your own farm.
- Different types of invertebrates.
- The arthropods.
- Types of fish scales.
- Turtle shells.
- Collect different types of bird feathers and identify the birds they came from.
- Form bird beaks from clay. Show how beak shape is adapted to the food each bird eats.

- Collect discarded birds' nests.
- Using plaster of paris, make and paint several different life-size birds' eggs.
- Assemble pictures showing different ecosystems. List plant and animal populations in each.
- Common compounds in our environment (including chemical formulas).
- Collect pictures of fossil fuels and by-products of fossil fuels.
- Display household appliances and their wattage (can also include cost of operation per hour).
- Biodegradable and nonbiodegradable materials.
- Collect and label pictures of clouds.
- Display weather maps from newspapers, showing air masses.
- Make replicas of three types of galaxies.
- Draw and label the parts of the heart from several animals.
- Assemble a collection of mollusks.

Apparatus

- Microscopes: a magnificent invention.
- Telescopes: pathways to the stars.
- How a generator produces energy.
- How to construct a wet cell.
- Build your own solar heater or cooker.
- Types of solar heaters.
- Turbines and their use in producing energy.
- Galileo and the telescope.
- What is a radio telescope?
- Using a weather map of the United States, choose major cities and chart the weather for each city, using appropriate weather symbols.

Grades 6 to 7

Experiments

- Can nonliving things grow?
- Effects of light on plants.
- Effects of root position on plant growth.
- Factors affecting germination.
- Will bean stems grow downward if the only light source comes from below?

- Testing acids and bases to determine their pH.
- Use red-cabbage juice to determine whether materials are acids or bases.
- Factors affecting leaf decay.
- Effects of colored lightbulbs on plant growth.
- Factors affecting wave frequency.
- The composition of soils in your area.
- Ways to desalinate salt water.
- At what temperature does condensation start?
- The relationship of relative humidity and barometric readings to changes in the weather.
- The effect of repetition on reaction times of different animals.
- Factors affecting the ability to memorize.
- Effects of heat on sugar.
- Effects of light sources in producing shadows.
- Effects of colored filters on fading.
- Effects of different types of fertilizer (artificial and natural) on plant growth.
- Effects of the depth of a planted seed on the plant growth.
- Effects of salt water and other liquids on plant growth.
- The behavior of mealworms.
- Preferred materials for nest building.
- Effects of different types of practice on learning rates.

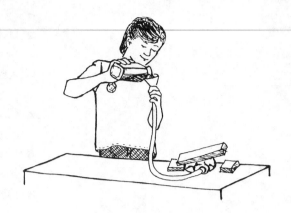

Demonstrations

- Use an art medium to show the main parts of an animal cell and a plant cell.
- Cell reproduction: a model of mitosis.
- Create replicas of one-celled animals.
- Prepare models of the different types of body tissue.
- Make models of birds' feet. Explain how they help each type of bird.
- Make plaster casts of horse and dog skulls, demonstrating the similarities and differences.
- Depict a major biome using a diorama or mural.
- Prepare a model of a world map and show the major biomes.
- The freshwater food chain.
- Use a prism and slide projector to produce a spectrum.
- How our eyes distinguish color.
- Effects of concave and convex lenses on light waves.
- Construct a working dimmer switch.
- Make an electrical question board.

- Make your own telephone receiver.
- Demonstrate how a television screen produces moving images.
- North American mineral resources.
- Fossil fuel deposits in North America.
- Demonstrate how the continents could have been joined together in a single land mass many years ago.
- Make models of the different types of mountains.
- Construct models of different types of satellites.
- Make a model of a rocket.
- How a rocket moves.
- Prepare a time line illustrating the U.S. space program.
- Construct a model of the brain showing the areas that control specific body functions.
- Using an outline of the human body, construct a replica of the human nervous system.
- Create the world's most successfully adapted imaginary animal. Explain how each adaptation would help it survive and prosper.
- Major nerve pathways in the body.
- How is paper made? How can it be recycled?
- Set up a project in which participants determine whether they are right-brain dominant or left-brain dominant.
- What colors make up sunlight?
- Assemble the bones of a chicken (make sure they are boiled clean before starting).
- Raise your own brine shrimp and report on their growth.

Research

- The use of plants in the world around us.
- The parts of a cell.
- What are single-celled organisms?
- Tissues, organs, and systems in the human body.
- Types of cells.
- The giant sequoias.
- The field of dendrochronology.
- Tropism and phototropism.
- Adaptations of seeds.
- Biological clocks and how they function.
- Migration patterns of selected birds.
- Structural adaptations of animals.
- Beavers: all you ever wanted to know.

- The primates: How they evolved.
- Evolution of reptiles.
- What happens in the body of a hibernating animal (temperature, digestion, brain activity)?
- Learned versus instinctive behaviors.
- Imprinting.
- How the human body reacts to exercise and inactivity.
- What is a biome?
- Exotic animals and plants.
- What are epiphytes?
- Plants and animals of the desert.
- Inhabitants of the ocean depths.
- Effects of human activities on animal biomes.
- A career in botany (or zoology or physics).
- The development of the atomic theory.
- The use of isotopes in medicine and industry.
- Elements and compounds used in industry.
- Physical and chemical changes in our environment.
- Nuclear medicine.
- How different kinds of lightbulbs work.
- Effects of a damaged ozone layer.
- How lasers are used in medicine and industry.
- How a telephone works.
- How can sound waves be recorded?
- The sending and receiving of radio/television signals.
- Is a career in chemistry for you?
- Role of forest fires in forest ecology.
- Renewable and nonrenewable resources.
- Benefits of recycling.
- Gasoline: from the ground to the station.
- Resources from the ocean.
- Sea floor nodules: what are they?
- Mariculture: is it feasible?
- Resources found in your own state.
- Changes in the Earth's crust.
- All about plate tectonics and continental drift.
- Drilling for oil: how and where?
- Earthquakes: earthshattering events.
- What causes volcanoes?
- How mountains are formed.

- The meteorologist's job.
- What is relative humidity?
- The gathering of weather data.
- The National Weather Service.
- Hurricanes: just a lot of wind?
- The Space Shuttle program.
- The positive and negative uses of satellites.
- Weather patterns on other planets.
- How microchips are designed and made.
- The function of the space probes.
- Could you be an astronaut?
- The future of the moon.
- Effects of space exploration on our lives on Earth.
- How industrial robots work.
- Functions of parts of the brain.
- The endocrine system: what does it do?
- All about biofeedback.
- How different organisms reproduce.
- Hereditary diseases.
- Location of volcanoes around the world.
- The origin of the moon.
- Chart the number of chromosomes in the body and sex cells of different plants and animals.
- Computer control in pattern weaving.

Collections

- Seeds and leaves from trees.
- Gather different sizes of firewood (before it is split). Count the annual rings to determine tree ages.
- Fruits and their seeds. Have participants match fruits with their respective seeds.
- Collection examples of animals that use protective coloration, protective resemblance, and mimicry.
- Different types of grains and their uses.
- Aquatic living things: marine and fresh water, plant and animal.
- Acids and bases used in everyday life.
- The inventions of Thomas Edison.
- Transparent, translucent, and opaque objects.
- Devices that have become smaller because of circuit technology.
- Recyclable materials.
- Items that use minerals.
- Display of nonorganic litter or trash found around the

school, the home, or in the neighborhood.

- Products made from fossil fuel feedstocks.
- Edible seaweeds.
- Pictures of objects in space.
- Make illustrations of the brains of different mammals.
- Objects that act like mirrors: plane, convex, or concave.
- Types of plant propagation (seeds, layering, sexual and asexual).
- Collect leaves and chart, compare, and contrast all their properties. Note similarities and differences.
- The phases of the moon (use photographs and diagrams).

Apparatus

- Use a microscope to view cells from different objects. Include slides. Diagram each cell.
- Perpetual motion machines.
- The spectroscope.
- Computers in the field of science.
- How to make a solar still.
- Seismographs: what are they?
- Weather instruments and the information they give us.
- Using a wet/dry bulb thermometer to determine relative humidity.
- Telescopes: the different kinds and their uses.
- How to build your own radio telescope.
- Make a working hygrometer.
- Microscope parts and their function.
- Make a pinhole camera.
- Create your own compass.
- The making of a turbine.
- A wind erosion recorder.

Grades 7 to 8

Experiments

- How water can be purified at home.
- Expansion rates of different metals.
- How sound is transmitted.
- Growth of rats in a vitamin-deficient environment.

- How copper plating takes place.
- Reaction of protozoa to different chemicals.
- Developing photographic film.
- Photosynthesis in lower species of animals.
- How gerbils learn their way through a maze.
- The growth of bacteria in different commercial disinfectants.
- How to measure the Earth.
- Chick hatching rates at different levels of humidity.
- Mold growth on different types of bread (wheat, white, rye).
- Growth patterns of yeast.
- The growth of grasses in different soils.
- How does acid rain affect seed germination?
- How we see colors.
- How the human digestive system works.
- An examination of plant cells.
- Reactions of seeds to different chemicals.
- Effects of salt versus sugar on plant or animal growth.
- How much water do different soils hold?
- What type of insulation holds in heat the best?
- Effects of junk food on gerbils.
- Does affection influence growth rate in hamsters?
- Effects on germination rates of seeds exposed to ultraviolet light.
- Effects of phototropism on different plants.
- Effects of light direction on plant growth.
- Effects of electricity on plant growth.
- Effects of cigarette smoke on selected insects.
- Effects of aspirin on the growth of selected plants.
- Do different types of music affect individuals' learning power?
- Effects of car exhaust on different plants.
- Ways to slow down plant growth.
- Effects of different colors on the eating habits of chicks.
- Effects of toothpaste on bacteria growth.
- Effects of noise on plant growth.

Demonstrations

- Wavelengths of sound produced by different musical instruments.
- Effects of light on the activities of rats.
- Growth and development of a chicken.

- A study of hydroponics.
- A panorama of optical illusions.
- Osmosis in plants: How does it work?
- What causes skin to darken in the sun?
- How pianos work.
- Effects of smoking on human health.
- How birds fly.
- How hair grows. Why hair falls out.
- Different types and degrees of noise pollution.
- How books or magazines are made.
- How candles are made: then and now.
- Things people write on.
- Effects of alcoholism.
- The birth of an island.
- The water cycle and how it operates.
- Mummies through the ages.
- Fluorescent light and seed growth.
- All about fingerprints.
- How does a rocket function?
- Pulleys and how they work.
- Different examples of water pollution.
- Contamination in our drinking water.
- Rocks: the ancient time machines.
- Plants in cooler environments.
- Biorhythms: do they affect our lives?
- Effects of colored light on goldfish.

Research

- How glass is made and used.
- Dinosaur extinction: how and why?
- How is fingerprinting done?
- Teeth: how to use them, how to take care of them.
- Famous caves.
- Erosion: its principles, causes, and cures.
- Cancer through the ages.
- The aerodynamics of different flying objects.
- The future of solar energy.
- Nuclear power: friend or foe?
- How hormones work in the human body.
- Different ways to dispose of garbage and litter.
- A history of photography.
- How plants and animals depend on each other.
- The Ice Age and its aftermath.
- Tsunamis.

- The beginnings of agriculture.
- How to turn a desert into farmland and vice versa.
- Laser beam technology.
- Archeology as a profession.
- The development of telephones.
- How electricity is made.
- Radioactivity: problem or potential?
- How babies develop: stages of gestation.
- The human brain: form and function.
- Causes and effects of acid rain.
- How streams and rivers get polluted.
- Effects of drinking and driving.
- How age affects memory.
- Is eye color related to vision?

Collections

- Samples of local soils.
- Homemade crystals in different solutions.
- Seashells from near and far.
- Preserved snowflakes: no two alike?
- Samples of different spider webs.
- Food types showing sources of different vitamins and minerals.
- Different types of electromagnets and their uses.
- Computers: past, present, and future.
- Stringed instruments and their sounds.
- Precious gems and where they're found.
- Local fossils.
- Engines, big and small.
- Sources of drugs.
- Drugs used in medicine.
- Plant foods versus animal foods.
- Pollutants in our everyday environment.
- Wild berries and nuts.

Apparatus

- Different types of pendulums.
- Thermometers, big and small.
- Uses of strobe lights.
- Robots.
- Crystal radios from yesteryear.
- Steam engines that made a difference.
- What a seismograph does.
- How a camera works inside and out.

- Different kinds of motors.
- Rockets into space.
- Simple machines used everyday in the home.
- Instruments used to study diseases.

In all, students have a wide range of possibilities from which to choose. Students should not feel restricted to these lists but can be encouraged to seek potential topics beyond the limits of these pages. The science world has unlimited areas to investigate: students should feel free to explore in each and every corner, guided by their interests.

9 Conducting Research

Before they begin any science project, students will find a wide variety of print and nonprint materials that can provide background information and offer additional ideas for exploration or development. Students should be encouraged to investigate as many different sources as possible to ensure a thorough understanding of their chosen topic.

The school library is a natural place to begin, but students should also explore the local public library, a nearby college or university library (many of which are open to the public), government agencies (which can provide needed materials free of charge or for a nominal fee), a local scientific laboratory (check in the phone book), newspaper or magazine offices, city or county agencies, mail order businesses that distribute science materials (see the Resources list).

 For more research ideas, refer to Web Resources in Chapter 15. Look for the section titled Research.

Materials to Investigate

Here are some materials students may wish to use.

encyclopedias
dictionaries
biographical dictionaries
atlases
pamphlets
records
newspaper files

maps
bibliographies
library card catalogs
professional indexes
almanacs
textbooks
graphs

brochures
magazines and professional
   journals
historical stories
photographs and art
charts
magazine indexes
public documents

Places to Go

Students often confine their research to the school library. It is certainly a good place to start—but only as a start. It should not be the sole source of information. Students should be encouraged to check out not only other libraries but businesses, government agencies, and the like. These investigations provide students with a well-rounded approach to their project: a vital concern of scientists everywhere. Here is a selection of places beyond the school library to explore.

college library
museums
scientific societies
local laboratory
historical society
recreation area
pumping station
state agencies
city/county offices
ranger station
park
Chamber of Commerce
waste treatment plant
TV/radio station
planetarium
manufacturing plants
hardware store
refuse collection firm
medical laboratory
zoo
TV specials
government publications
   (check Resources list)

botanical garden
gardening center
farm
state gamelands
state seashore areas
newspaper office
college science department
supermarkets
restaurants
glaziers
pet store
arboretum
florist or plant nursery
food-processing plant
university laboratory
wildlife preserves
travel agency brochures
science periodicals
   (check Resources list)
publications of
   professional unions
computer databases

People to See

Students need to be aware that a successful science fair project may require consultations with individuals other than their teachers or parents. By interviewing experts in the chosen subject area or talking over the project with them, students may obtain new data or additional insights to incorporate into the project plan. Such discussions offer students an opportunity to share ideas and discuss diverse aspects of a chosen topic. People are often eager and honored to talk with students about their projects and can be a most valuable resource during the investigation process. Here are some individuals who might be interviewed (an interview sheet appears at the end of this section).

science teachers
professors
electricians
friends and neighbors
park rangers
college students
plumbers
librarians
veterinarians
city or council government
    officials
astronomers
high school students
computer operators
musicians
zoologists
gardners
zoo personnel
meteorologists

environmentalists
ecologists
people encountered
    during field trips
scientists
corporate and research
    librarians
science writers
factory workers
doctors, nurses
farmers
medical lab workers
construction workers
travelers
sanitation workers
biologists
cartographers
conservationists
cooks

Interview Sheet

Name of person being consulted: Position:

_____

Phone number for further questions: E-mail address:

_____

Date of interview: Time of interview:

_____

Location of interview:

_____

Questions and Answers:

1. _____

_____

_____

_____

2. _____

_____

_____

_____

General Comments/Reactions/Ideas to Pursue:

_____

_____

_____

_____

_____

_____

10 The Scientific Method

Many students elect to conduct an experiment for their science fair project. An experiment allows a student to investigate an area of science by means of principles and methods scientists use every day. An experiment should be effectively designed so that the student can discover the answer to a precisely defined problem. Students often make the mistake of selecting a problem that is too general or too broad in scope—one they do not have the resources, materials, or time to investigate properly. Most students need guidance so that the problem they choose to investigate is well within their capabilities and for which appropriate resources are available.

The scientific method consists of a series of steps that must be followed to ensure an effectively designed experiment. Note that the steps allow some leeway, offering students many ways to examine and explore an area of interest. Nevertheless, if the project is to yield scientific data that will expand the students understanding, each of the following steps is essential.

Steps of the Scientific Method

1. Identify the problem.
2. Refer to authoritative sources.
3. Ask an appropriate question.
4. Develop a hypothesis.
5. Conduct experiments.
6. Keep detailed records of methods and results.
7. Report the experiments.
8. Analyze the results.
9. Develop a conclusion.

Identifying the Problem

With so many potential topics from which to choose, students must narrow their choices to a specific one. Here is where guidance from teachers and parents becomes so important. It is not unusual for students to decide to do an experimental project on a broad topic like "grass" or "white mice," for example. Here is where students should be asked some questions that will assist them in defining a more specific problem. For example: "What feature about grass interests you the most?" "What question would you like to ask a gardener about the growth rate of different grasses?" "How do you think mice eat their food?" "How do mice survive cold temperatures?" Assist your students in carefully defining the scope of a problem for investigation and narrowing it to a level they can explore. This is a process scientists go through regularly and one that is essential to a well-designed, successful experiment.

Referring to Authority

Long before scientists begin to set up their experiments, they conduct some research in their chosen area. This means reading books, magazine articles, pamphlets, brochures, or any other printed information concerning their topic. It also means talking with or obtaining information from experts in the field. This is done through telephone conversations, personal visits, or attendance at special meetings or conferences. Like scientists, students should be prepared to conduct some investigative research before initiating an experiment. These discoveries can yield a significant amount of valuable data that sharpens a student's understanding of a selected field.

Asking an Appropriate Question

To develop an effective science experiment, students need to formulate a very specific question about the chosen area of interest. Parents and teachers can help. For example, a student who has an interest in learning about how plants grow in different colored lights might ask: What is the difference in the rate of growth of four different plants, each grown in a different-colored light? Or a student who is interested in the nutritional needs of mice might ask: What is the difference in the weight of mice raised on a diet of junk food in comparison to that of mice raised on a nutritious diet? Notice how each question is very specific; it indicates the subjects to be studied, and the variables that will be observed. Being specific and identifying variables is important in helping the student sharpen his or her focus and carefully define the area to be investigated.

Developing a Hypothesis

After students have designed an appropriate question, they must turn that question into a hypothesis. A **hypothesis** is an educated guess, a statement of how the scientist thinks the experiment will turn out. It is a prediction, based on the best available information, of what the scientist believes will happen at the conclusion of the experiment. Although the hypothesis is founded on factual data the student has collected during the research stage, it is the student's *opinion* deduced from those facts. A well-constructed hypothesis identifies the subjects of the experiment (plants, mice) and states what is being measured (rate of growth, weight), the conditions of the experiment (different-colored light sources, junk food verses regular food), and the results expected (light colors produce faster growth rates than dark colors; a nutritious diet produces higher weights than a junk food diet). Thus a student's question about a specific area of interest can be developed into a hypothesis that forms the foundation of the student's investigation. For example:

- Bean plants grown under dark-colored light will grow more slowly than bean plants grown under light-colored light because of a lack of sufficient ultraviolet light waves.

- Mice raised on a diet of junk food will show lower body weights after 6 weeks than mice raised on a regular diet because of a deficiency of necessary nutrients in the junk food.

Conducting the Experiment

Testing one's hypothesis is at the heart of the scientific method. It is here that the student will organize and conduct an investigation examining the effects of changes in certain experimental conditions or experimental factors. In short, the student will learn what happens when a condition is created or altered. In addition, the implications of that change are also explored.

It is important that the student test or examine *one idea at a time*. It is easy for students to expand their experiments far beyond the limits established with the original hypothesis; however, for any results to be valid, students must adhere to the original design. Often, this means a process of trial and error in which a problem is approached from many angles before a hypothesis can be confirmed.

It is also at this stage that students must decide how many times they will conduct their experiment, the number of subjects or items being subjected to the test, how long it will last, and what special materials they will need. Students must also decide how they will be measuring the effects of the experiment. Will it be done over a period of time, include a variety of sample types, or measure height, weight, growth rate, heartbeat, or other variables? For example, will the experiment take place over a few hours, days, or weeks? Will it include different varieties of animals or plants? What measuring instruments (scales, balances, stopwatches, clocks, thermometers, anemometers, wind gauges, yardsticks) will be used? Will the experiment be conducted in daylight or darkness, at high elevations or low elevations, in a lab or the family basement? The conditions under which the experiment will be carried out must be clarified prior to as well as during the entire experiment.

Keeping Records

Scientists always maintain records of everything they do during the course of an experiment. Students, too, should be encouraged to keep a log or record book of what they do and observe during the course of the investigation. Such record keeping permits the student to keep track of the individual events of the experiment and it provides a reference of identifying any errors that may creep into the experiment.

Not only does the log or record book provide an accurate summary of "events" of an experiment; it is also important in showing others how the experiment was carried out. Others should be able to duplicate the experiment simply by following the student's record. Indeed, the well-maintained notebook is an invaluable part of any science fair experiment because it details the steps and procedures the student went though to arrive at new information.

Repeating the Experiment

Conducting an experiment once usually does not provide the scientist with sufficient data upon which to base a decision or conclusion. Thus it is important that the student plan adequate time to conduct the experiment more than once. Such a practice ensures that the results obtained the first time are accurate and precise. It also guarantees that any conclusions drawn from the results of the experiment are based on a wealth of information and not just a few isolated details.

Analyzing the Results

After the experiment has been conducted and all the necessary data collected, it is time to analyze that information. What facts, numbers, or statistics were produced as a result of the experiment? Did three of the plants show slower rates of growth than a fourth? Was there a difference between all individual plants? How many mice had lower body weights on the junk food diet? Did the growth patterns differ significantly between the two groups of animals? The collection of this information and its analysis are vitally important parts of the entire project. It is here that the student assembles and looks over the available results in order to begin formulating a conclusion.

It is important to point out to the student that the data gathered may *not* confirm the original hypothesis. That is, as a result of this experiment the student may discover, for example, that there is no difference in the growth rate of

bean plants grown under different colors of light. That's OK.
Students need to understand that their original hypothesis
was simply an educated guess based upon information at
their disposal at the start of the experiment. It is possible
that the results will not confirm that hypothesis but rather
refute it. This happens to scientists all the time and is a
natural and normal part of the scientific method. Science
is advanced just as much by the knowledge that light
color is not a factor in growth as by a finding that it is
a factor. In an experiment, success is neither a positive
or a negative finding but any clearly substantiated,
repeatable result.

Developing a Conclusion

Now that the student has conducted the experiment,
collected the necessary data, and analyzed the results, it is
time to formulate a conclusion. The conclusion should
provide some answer to the original question (see above),
although it is entirely possible that the experiment was
unsuccessful in proving the hypothesis. There is certainly
nothing wrong with a conclusion indicating that the
question still remains unanswered. The importance of the
experience lies in the student's having an opportunity to
investigate and learn about an area of interest by means of
the scientific method. Arriving at an ideal conclusion is
not the goal—wrestling with the problem is.

In a sense, the conclusion represents what the student
actually learned by conducting the experiment. It is also
an opportunity for the student to suggest needed
improvements in the experimental design or changes that
could be made in attempting the experiment in the future.
Most important, the conclusion should contain a
statement or series of statements by the student on the
importance of a well-rounded and nutritious diet in the
maintenance of proper body weights for animals or
humans. The conclusion, then, is an opportunity for the
student to draw relationships between the experiment and
the world in which he or she lives.

11 Presenting the Project

After students have selected a topic, investigated or experimented with that topic, and come to some conclusions about that area of science, they will want to display their efforts for the science fair. Presenting the project can be one of the most satisfying parts of the entire experience. The science fair project display is the culmination of weeks of study and preparation: It's here that students can demonstrate their ingenuity and creativity in sharing what they have learned.

Each project entered in a science fair must consist of three elements: the display unit, the exhibit materials, and the written report. Together, these elements present a complete and through examination of an area of interest, a collection of new knowledge, or the results of a self-initiated experiment. In most science fairs, the displays are evaluated and thus must present a complete picture of the student's efforts for judges and observers alike.

Most science fairs provide students with display tables on which the project can be set up. *Before* designing the display, find out the length, width, and height of the tables to make sure there is enough room to arrange the display and that any written information can be read easily by observers. Keep in mind, however, that not all projects will need a display table (see illustrations throughout this book).

Display Backdrop

The display unit (also known as the backdrop) is crucial to presentation: It is what people see first; it establishes the "professionalism" of the student's efforts. As a kind of advertisement for the project, it must be well constructed and designed for maximum visual effect.

Copyright © Good Year Books

59

Materials

A good display unit must be constructed of sturdy and durable materials. It must stand for several days, so strong, rigid materials are preferred.

Pegboard

Pegboard has the advantage of providing pre-drilled holes from which displayed items it can be hung. Easily available at most lumberyards or hardware stores, it can be cut to any size or shape. Its only disadvantage is its thinness and its tendency to buckle when heavy weights are hung from it. This can be corrected by nailing strips of wood around the perimeter of the back of the pegboard to provide the necessary support.

Plywood

Plywood comes in many thicknesses ($\frac{1}{4}$", $\frac{1}{2}$", and $\frac{3}{4}$" are all commonly available). It provides a more than adequate backboard for any display unit. It can be cut into sections and hinged together for an effective display. If the plywood is to be painted, you'll need to buy "A" or "B" grade wood. If however, the plywood will be covered with other materials such as felt, construction paper, or foil, "C" or "D" grade plywood will be adequate.

Corkboard

Corkboard is a lightweight material available in variety stores and lumber yards. It is not as rigid as other materials but has the distinct advantage of being easily portable from home to school.

Particle Board

This extremely sturdy material is obtainable at any lumber yard or building supply store. It provides a very solid backdrop for a display unit, particularly for backboards displaying heavy objects. Its disadvantages are its weight and difficulty in cutting.

Foam Board

This is an ideal material for display units and can be found in art supply or graphics supply stores. It is extremely lightweight, comes in thicknesses up to 8inches, and can be very easily cut with a utility or razor-blade knife. It consists of plastic foam covered on both sides with paper. Another advantage, aside from its light weight and sturdiness, is that when scored on the back with a knife it can be shaped into a curved display unit. Additionally, its surface can be easily painted and materials can be fastened to it with cellophane tape, pins, or glue.

Undesirable Materials

Materials that should not be used for a display unit include cardboard and posterboard. These materials are too light to stand on their own or hold display items for long periods.

Construction

The display unit must be freestanding, that is, it must stand on its own for several days. Thus, most display units will consist of three equal-sized panels hinged together. Display units consisting of two or four panels are less frequently used. If foam board is chosen, it is possible to construct a curved display unit—a technique that adds a distinctive touch to the project. Display units come in many sizes, but here are some suggested dimensions based on standard measurements (in inches) of materials commonly used in science fair displays:

Height		Width		Height
60"	*	48"	*	30"
48"	*	48"	*	36"
60"	*	48"	*	36"
48"	*	36"	*	30"

Setup

To provide for an attractive display unit, consider painting the backdrop or covering it with construction paper, adhesive backed paper, wallpaper, burlap, or some other appropriate material. It is best to keep the colors neutral so as not to detract from the display itself. The completed backdrop will include information about the project as a whole as well as other data summarizing important parts of the investigation.

The setup of a display unit is vitally important. Students should be prepared to make their backdrops attractive without including so much material that the display becomes visually crowded. By experimenting with different designs and formats, students should be able to come up with a mix that will enhance and illustrate the information.

The following information should be included on the display unit (see illustrations throughout this book for examples of different arrangements).

Purpose

This statement lists the student's reasons for pursuing the project. What did the student hope to learn by investigating this area?

Procedure

What did the student do to carry out his or her plan of action? What methods or materials were used to discover new information about a topic?

Problem

A problem statement outlines a condition or fact the student is uncomfortable with or seeks to investigate. *Note:* A problem statement is most commonly included in experiment-type projects.

Hypothesis

A hypothesis is an educated guess or prediction about what the student thinks will happen. Note: Hypotheses are used mainly in experimental projects.

Title of the Project

The title must describe, very succinctly, the focus of the project. It should be short (10 words or less), neatly lettered, and easy to read.

Results

What did the student learn during or after his or her investigation? In other words, what facts were discovered that were not known before?

Conclusion

This statement summarizes the student's investigation. It should offer an answer to the student's original questions. Students may discover something not originally planned—that too should be included.

Visual Aids

These include photographs, charts, surveys, graphs, data, drawings or paintings, diagrams, or other illustrative materials that enumerate vital information gathered during the project.

Lettering

An important part of any display unit is the lettering. Good lettering can add much to a display by conveying an important message to all who look at it. It is important, therefore, that the lettering and signs used on the display unit be neat and of proper size. The title should have the largest letters, while signs posted over each supplemental section are smaller. Good hand-lettering is sometimes sufficient, but stencils or press-on letters (available at any art supply or large variety store) add a true professional appearance to the entire display. If available, computer graphics programs can be used to create labels, titles, and signs. The student should check and double-check all spelling and punctuation.

Exhibit Materials

The materials, items, devices, and samples shown in front of the backdrop unit can be an exciting part of any science project. These materials should reflect the items used throughout the student's investigation; they should provide a firsthand look at the scope of the project. In projects displaying collections, a good cross section of different types or groups of the selected items would be displayed. In a project illustrating an apparatus or group of related apparatuses, examples of those items would be put on display for the viewer. If the student chooses to do an experiment, then the materials used throughout the experiment are set up for viewing. As a rule of thumb, the display items should tell a story or illustrate a concept sufficiently so that the student scientist need not be present to explain the entire project to an observer. Here are some procedures to keep in mind in setting up this part of the project.

- **Safety first!** Exhibit items should present no hazards to observers who may view the display. Thus, no breakable or dangerous items should be included. If electricity is used, safeguards must be observed to prevent electrical shocks or hazards. (Battery-powered equipment is preferable.)

Safety Considerations

The health, safety, and well-being of students should be paramount in any scientific investigation or science fair project. Share these safety guidelines with students and parents. Plan time to discuss them with students early in the planning stage of the science fair.

❑ Do not incorporate supplies or equipment into science fair projects until you have been given specific instructions in their use.

❑ Do not use dangerous science materials, such as chemicals or sharp instruments, in any science fair project.

❑ Do not include activities that involve participants touching, tasting, or inhaling unknown materials.

❑ Do not include any live animals in any science fair project.

(more)

❏ Wear eye-protection devices for activities that pose a potential risk to eye safety.

❏ Projects that involve the use of heat or electricity should be approved and examined by an adult before being submitted to a science fair.

❏ Check electrical wiring on any equipment for frayed insulation, exposed wires, and loose connections.

❏ Make sure the display is sturdy and will not fall or collapse if bumped.

❏ Ensure safe access to all parts and all areas of a science fair project. Consider safe and easy access to a display by physically handicapped individuals.

❏ Ensure that small items cannot roll or fall off the display table. Take special care with any items that could be ingested by small children.

❏ Make sure the display is safe even if left unattended.

Science fair organizers should consider these additional points:

❏ Make sure lighting is sufficient to ensure that displays are properly lit and to avoid dark spots or hazardous areas.

❏ Check, double-check, and triple-check every project and every project component to ensure that they are safe for the students who created them and all viewers.

- **Set up an attractive format.** Have the students experiment with a variety of designs and formats to arrive at the most visually appealing one.

- **Avoid clutter.** It is important to include enough items to illustrate important concepts of the project, but it is equally important to avoid crowding the table. Too many items detract from the display just as much as too few.

- **Avoid using liquids or chemicals.** The use of water, chemicals, or other liquids is discouraged, particularly for displays that will be standing for

several days. Any spillage could be a hazard to the display or a neighboring display. It would be preferable to take photographs of selected vessels or liquids at home and then post the photos on the display unit. (Allow time for doing this.)

- **Seal in smelly items.** If molds or decaying items are to be exhibited, they must be sealed tightly inside glass or plastic jars. All cautions must be observed to prevent these materials from being released into the surrounding area.

- **No animals of any kind should be included in a display, especially displays left overnight.** The care and maintainance of the animals cannot be guaranteed by the science fair director(s). The strain of having countless observers peek, point, and poke at animals can place an unnecessary strain on household pets. Thus, it is important that students carefully consider any projects that involve the use of animals. The animals should be left at home and photographs of them included as part of the display.

Written Report

The written report is a capsule summary of everything the student did to investigate the selected topic. It contains all the information the student collected or learned during the weeks leading up to the actual science fair. Whether the student decides to do an experiment, assemble a collection of objects, demonstrate a scientific principle, conduct some research into an interesting area of science, or show a particularly interesting piece of scientific apparatus, it will be necessary to record observations and information in written form. This written report provides observers with vital data on the scope of a project as well as its effect on a student's understanding of the topic.

Usually 5 to 30 pages in length, the report provides observers with a blow-by-blow account of everything the student did throughout the length and breadth of the project. It is meant to provide readers with a succinct, detailed account of the chosen project—including its impact on the student. Above all, it provides students with an opportunity to think about all the dimensions of their projects and to share their ideas with others.

Reports should be neatly bound in an attractive folder or binder (available at any variety, department, or stationery store). It is preferable for the report to be typewritten or done on a computer, but a neat handwritten copy may also be acceptable. Any written report for a science fair project should include:

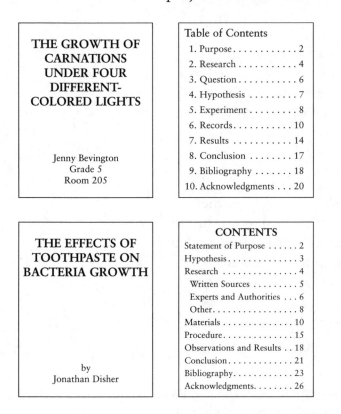

Title Page

The first page in the report should include the title of the project as well as the name and grade of the student.

Table of Contents

This page provides the reader with a list of the different parts of the project and the page number on which each section can be found.

Statement of Purpose

This two- or three-sentence statement explains what the student expected to discover by investigating the chosen topic. It also gives the reason why the student chose to learn more about that subject.

Hypothesis

Students who select an experiment to perform should include a hypothesis in the written report. A hypothesis is an educated guess about what the student thinks will occur as a result from conducting the selected experiment. It is not necessary to include a hypothesis for other types of projects.

Research

This is the part of the report that contains all the background information the student collected about the chosen topic. Any books or articles read, authorities consulted, or outside materials collected should be written in the student's own words and not copied from an encyclopedia or other reference.

Materials

This is a list of all the materials and supplies used in the project. Quantities and amounts of each should also be indicated, especially if the student conducted an experiment.

Procedure

Here the student lists and describes steps he or she undertook to complete the project. Usually presented in a numbered format, this part of the report shows the stages of the project in such a way that others can reproduce the procedure.

Observations and Results

Here the student tells what he or she learned from the project. What new information was provided as a result of pursuing this topic? What does the student know now that wasn't known before? It is important to include any graphs, charts, or other visual data that summarize the results of a study.

Conclusion

This is a brief statement explaining why a project turned out the way it did. Students should explain why the events they observed occurred. Using the word *because* is a good way to turn an observation into a conclusion. If an experiment was chosen, the conclusion should tell whether the hypothesis was proven or not proven.

Bibliography

The bibliography should list all the printed materials the students consulted in carrying out the project. Items should be listed in alphabetical order in a standard format. The resources are one example of a format.

Acknowledgments

Here the student thanks all the individuals who assisted in the research or development of the project (including Mom and Dad). Everyone the student interviewed, including teachers, scientists, and other experts in the field, should be mentioned here.

12 Judging Projects

Evaluating science fair projects can be a difficult task for students as well as judges. Students who plan to enter projects deserve guidelines for their efforts—not to compete with other entrants but to make the current entry as good as possible and to prepare for future entries.

The judging criteria listed in this chapter have been garnered from science fairs held around the country. They provide teachers, parents, and, most important, students with significant criteria against which to gauge projects. They are appropriate to evaluate all kinds of entries submitted to a science fair, but they are also designed to serve as a self-evaluative tool prior to entering one's project in the fair. These guidelines ensure that all students are being evaluated according to an established set of criteria and that teachers and parents are aware of the elements that constitute a well-planned display.

What Judges Look For

Judging science fairs projects can be very subjective. Judges and the criteria they use can vary from one science fair to another. Despite these variances, judges typically consider five major factors when evaluating projects: creativity, scientific thought, thoroughness, skill, and clarity. Each of these factors is discussed briefly below. The evaluation checklist that follows provides specific criteria for each one.

Creativity

Judges appreciate projects that are unique. When there are dozens of solar system models or erupting volcanoes in an exhibit area, judges tend to give them lower scores. Ingenuity, uniqueness, and creativity most often distinguish a great project from a good one. In short, the more original a project is, the more points it will garner from the judges.

Scientific Thought

Science fair judges favor projects that demonstrate an understanding of scientific principles and a comprehension of the scientific method in solving a problem. Judges look for a clearly defined hypothesis, a clear statement of the problem, research tailored to the complexity of the problem, clear and organized experimental procedures, sufficient data collection, and appropriate data presentation.

Thoroughness

When judges examine a project close-up and then stand back to observe it as a whole, they are looking for a comprehensive and inclusive presentation. Does the project tell a story or is it merely a collection of random facts? Was everything included or is important information missing?

Skill

Most judges have examined and evaluated numerous science fair displays over the years. They are mindful of whether projects have been constructed by students or by adults. The elaborateness, sophistication, and complexity of a display provide hints about who did most of the work. While adult supervision is encouraged, projects must clearly represent the work of individual students. Those that do typically receive high marks.

Clarity

Judges consider whether an overall display is visually appealing, properly prepared, and clearly presented. The display should evidence a complete idea that was planned and executed over an extended period of time rather than hastily assembled at the last minute. The sturdiness of the display, the completeness of any written materials, and the presentation of exhibit materials are all significant elements in this part of the evaluation.

How to Use This Checklist

The major parts of any science fair project are the *display* and the *written report*. Each has been assigned a possible 100 points. In addition, projects involving experiments need to be evaluated for use of the scientific method.

To help your evaluation, points have been assigned to the major components of each category, and some criteria are listed below each. *Not every item in each list will apply to every project.* Check off the applicable ones that you believe have been fulfilled successfully, then assign a value between zero and the total potential points and fill it into the blank beside the component. Add the points, and you'll have a standard for comparing your project, or a project you are judging, with the best possible results.

Display

Score

_____

Creativity (30 points)

❏ Are the materials presented imaginatively?
❏ Is there a distinctive approach to problem solving?
❏ Is the project or display original?
❏ Is the display designed in an unusual way?
❏ Is there a variety of equipment or items?
❏ Is the project out of the ordinary?
❏ Is new and interesting information included in the display?
❏ Are the data or results interpreted appropriately?
❏ Has the student shown inventiveness?

Scientific Thought (30 points)

- ❑ Is the experiment designed to answer a question?
- ❑ Are the procedures appropriate to the area of investigation?
- ❑ Is the topic or problem stated clearly and completely?
- ❑ Has scientific literature been sited?
- ❑ Have scientists or other experts been consulted?
- ❑ Has a systematic plan of action been stated?
- ❑ Is there a need for further research or investigation?
- ❑ Is there an adequate conclusion?
- ❑ Is a project notebook provided with the display?
- ❑ Is the project notebook sufficiently detailed in relation to the scope of the project?
- ❑ Have any problems or limitations that occurred been noted?
- ❑ Is the amount of data commensurate with the scope of the project?
- ❑ Does the student understand all the facts and/or theories?

Thoroughness (15 points)

- ❑ Is the project complete?
- ❑ Does the project represent a sufficient amount of time?
- ❑ Is a problem adequately answered or pursued?
- ❑ Are the notes complete?
- ❑ Are other potential solutions or approaches indicated?
- ❑ Does the project include a display unit, three-dimensional items, and a written report?
- ❑ Does the project tell a complete story?
- ❑ Is the information complete?
- ❑ Were all potential sources of information consulted?
- ❑ Is the display lightweight and portable?
- ❑ Is a sufficient number of items included in the display?
- ❑ Is the conclusion supported by results from an experiment?

Skill (15 points)

- ❑ Does the project represent the student's own work?
- ❑ Does the project represent quality workmanship?
- ❑ How much outside assistance did the student need?
- ❑ Is the project artistically pleasing?

❏ Does the student know the subject well?
❏ Is there anything dangerous about the display?
❏ Did the display take a lot of time to set up?
❏ Was assistance necessary in setting up or preparing the display?
❏ Does the project indicate extensive planning?
❏ Is all equipment used within the student's level of understanding or expertise?
❏ How much supervision did the student require?

Score
_____

Clarity (10 points)

❏ Are titles and written descriptions neat, legible, and large enough to read?
❏ Are the data clearly presented?
❏ Can the average person understand the project?
❏ Is the written material well prepared?
❏ Is the project self-explanatory?
❏ Are drawings and diagrams neat and attractive?
❏ Is the presentation logical and sequential?
❏ Is every piece of material important to the display?
❏ Is the display colorful and attractive?
❏ Are any supplemental guides provided?
❏ Are discussions clear and straightforward?

Total for Display
(100 possible points)

_____

Written Report

Score
_____

Title Page (2 points)

❏ Is it present?
❏ Are the title of the project and the student's name included?

Table of Contents (5 points)

❏ Are all parts listed?
❏ Are all sections listed in order?
❏ Are page numbers listed and correct?

Statement of Purpose (15 points) _____

❑ Does it pose a question that can be investigated
   or measured?
❑ Does it pertain to the experiment or project
   conducted?
❑ Is it logical and defendable?
❑ Is it clear and understandable?
❑ Is it within the student's ability level?

Hypothesis (Included for experiments only) No points

❑ Does it answer the purpose?
❑ Does it tell what the student is trying to prove
   with the project?
❑ Is it clear?
❑ Is it scientifically sound?

Research (15 points) _____

❑ Does the research pertain to the topic?
❑ Is it complete and through?
❑ Does it represent a diversity of sources?
❑ Is it representative of the student's ability?
❑ Have both print and nonprint sources been
   consulted?

Materials (10 points) _____

❑ Are all materials listed?
❑ Are specific amounts given?
❑ Are there sufficient materials?

Procedures (15 points) _____

❑ Are procedures listed in chronological order?
❑ Could the project/experiment be replicated?
❑ Are the procedures easy to follow?
❑ Are they in a logical order?

Observations (15 points)

- ❏ Do observations indicate what was done in the project?
- ❏ Did the student choose the best form for recording the observations?
- ❏ Are observations clearly labeled?
- ❏ Are they sequential?

Conclusion (15 points)

- ❏ Does it answer the purpose?
- ❏ If an experiment, does it adequately explain any results?
- ❏ Does it tie the entire paper together?
- ❏ Is it sufficient in form and length?

Bibliography (8 points)

- ❏ Is it in alphabetical order?
- ❏ Does it follow the required format?
- ❏ Is it sufficient in terms of the scope of the project?
- ❏ Have primary, scientific sources been consulted?
- ❏ Is range and scope of the bibliography reflected in the report itself?

Total for Written Report
(100 points possible) _____

Total for Project
(200 points possible) _____

Questions Judges Ask

During the judging process, many science fair judges like to talk with students about their projects. Typically, these are informal chats that don't require students to prepare notes or scripted speeches ahead of time. These conversations, however, provide judges with important information about the relationship between a student and his or her project.

Although the questions judges ask will vary depending on the specific elements and features of a particular project, participants often are asked the generic questions listed below. Teachers should share these questions with students before the initial setup of projects in the exhibit area. In addition, teachers may wish to conduct mock interviews with selected students to discuss appropriate responses.

- "What did you learn as a result of this project?"
- "If you could do the project again, what would you change?"
- "What was the most difficult part of your project?"
- "What was the least appealing part of your project?"
- "What would you want to share with others regarding your project?"
- "How long did you work on your project?"
- "What kind of help did you get for your project?"
- "What type of research did you have to do?"
- "Would you choose to do this project again if you knew the amount of work involved?"
- "How did your project help you learn more about science?"
- "If you had more time, what would you add to your project?"
- "What will you do now that your project is completed?"
- "Will you enter next year's science fair?"
- "What other kinds of projects would you like to do?"

 For more information about the judging process, refer to Web Resources in Chapter 15. Look for the section titled Judging Science Fairs.

13 Student's Planning Guide

TO THE STUDENT: A science fair project can be one of the most exciting projects you will ever do in school. But it takes time, planning, research, preparation, and lots of hard work. You will discover much about your chosen topic and much about yourself, too. You will examine, probe, and experiment with many ideas, techniques, and scientific principles, learning more about the world in which we live and more about the important work scientists do to help us understand our world. In short, you are about to begin one of the most thrilling journeys of your life—a journey into new discoveries and new adventures in science.

The Planning Guide on the following pages is designed to help you plan your journey both before and during your science fair project. Use it to select a topic for investigation, to decide what procedure to follow in exploring that topic, and to plan your steps for putting your completed project together for display at the science fair. Read this guide carefully and fill in or check off the necessary steps: It will make your work much easier. Remember that the success of your project will depend on the amount of work you wish to put into it before it is displayed at the science fair. Keep in mind that you can always ask other people (for example, your parents or teacher) for help or guidance in carrying out your project: Scientists help one another all the time! Now get ready—and **good luck!**

Science Fair Planning Guide

Things That Interest Me

1. _____

2. _____

3. _____

4. _____

5. _____

How much time can I spend on my project each week:

_____ hours

What area of science interests me the most:

❑ Life science
❑ Earth and space science
❑ Physical science

What type of project interests me the most:

❑ Experiment (using scientific method)
❑ Demonstration
❑ Research
❑ Collection
❑ Apparatus

What kind of science materials or equipment do I enjoy or am I familiar with?

_____

_____

_____

_____

_____

Possible Topics

1. _____

_____

Materials I already have: _____

_____

Materials I will have to buy: _____

_____

Help I will need with this topic:
None _____ Some _____ A lot _____

How difficult will this be for me?

Very difficult _____ Somewhat difficult _____ Easy _____

2. _____

_____

Materials I already have: _____

_____

Materials I will have to buy: _____

_____

Help I will need with this topic:
None _____ Some _____ A lot __

How difficult will this be for me?

Very difficult _____ Somewhat difficult _____ Easy _____

3. _____

_____

Materials I already have: _____

_____

Materials I will have to buy: _____

_____

Help I will need with this topic:
None _____ Some _____ A lot _____

How difficult will this be for me?

Very difficult _____ Somewhat difficult _____ Easy_____

Final Topic Choice: _____

Questions/Problems to Explore

Some questions about my topic I want to find answers to:

1. _____

2. _____

3. _____

4. _____

Conducting Research

Printed and audiovisual materials I should find and read:

1. _____

2. _____

3. _____

4. _____

5. _____

Places I could visit:

1. _____

2. _____

3. _____

4. _____

5. _____

People I could talk to:

1. _____

2. _____

3. _____

4. _____

5. _____

Preliminary Timetable

Here is what I plan to do each week (subject to change).
Students: Fill in 12 or 6 weeks according to your
scheduled opening date.

Week 1 (_____ to _____ **)**
a.
b.
c.
d.

Week 2 (_____ to _____ **)**
a.
b.
c.
d.

Week 3 (_____ to _____ **)**
a.
b.
c.
d.

Week 4 (_____ to _____ **)**
a.
b.
c.
d.

Week 5 (_____ to _____)
a.
b.
c.
d.

Week 6 (_____ to _____)
a.
b.
c.
d.

Week 7 (_____ to _____)
a.
b.
c.
d.

Week 8 (_____ to _____)
a.
b.
c.
d.

Week 9 (_____ to _____)
a.
b.
c.
d.

Week 10 (_____ to _____)
a.
b.
c.
d.

Week 11 (_____ to _____)
a.
b.
c.
d.

Week 12 (_____ to _____)
a.
b.
c.
d.

For Experiments Only

Here are the steps for the scientific method. Fill in each one as a guide for your experiment.

■ Problem I want to explore:

_____

■ References I will consult:
Printed and audiovisual:

_____

_____

Places

_____

People

_____

■ Specific question I will examine:

_____

■ My hypothesis is:

_____

■ The experiment will consist of
Subjects:

_____

Conditions:

_____

Tests:

_____

Time:

_____

Special materials:

_____

_____

Planned steps:

1. _____

2. _____

3. _____

4. _____

5. _____

6. _____

7. _____

8. _____

Record keeping
■ What I did during the experiment:

_____

_____

_____

_____

_____

■ Results I got when I repeated the experiment:

_____

_____

_____

_____

_____

■ Final results of my experiment:

1. _____

2. _____

3. _____

4. _____

5. _____

■ Conclusions I can base on those results:

_____

_____

_____

■ Did I prove my hypothesis?   ❏ Yes   ❏ No

_____

Why or why not?

_____

_____

_____

■ What further or different research can I suggest?

_____

_____

_____

_____

_____

_____

Presenting the Project

DISPLAY TABLE

■ Dimensions:

Height: _____ Width: _____ Depth: _____

STUDYING SOIL EROSION

■ Table cover materials needed:

1. _____

2. _____

3. _____

DISPLAY UNIT
■ Materials needed:

_____   _____

_____   _____

_____   _____

■ Dimensions of panels:

Height: _____ Width: _____ Depth: _____

■ Written data to be included on display (check each one):
❏ Purpose
❏ Procedure
❏ Problem
❏ Hypothesis (for experiments)
❏ Title
❏ Results
❏ Conclusion

■ Written data to be included on display:

_____

_____

_____

_____

_____

_____

■ Visual aids (check each one):
❏ Photos
❏ Charts
❏ Graphs
❏ Artwork
❏ Diagrams
❏ Cartoons
❏ Pamphlets
❏ Brochures
❏ Mural
❏ Magazine clipping(s)
❏ Newspaper clipping(s)
❏ Poster(s)
❏ Drawing(s)

■ Which of the following do I want to include:

❑ Apparatus

What types:

_____

_____

_____

❑ Specimens

What types:

_____

_____

_____

❑ Demonstration items

What types:

_____

_____

_____

❑ Special materials

What types:

_____

_____

_____

■ Are the items safe?

❑ Yes   ❑ No

■ If "no," what do I need to make them safe?

_____

_____

_____

■ Will observers be able to handle them?

❑ Yes   ❑ No

WRITTEN REPORT

■ Do I have an attractive folder?
❑ Yes   ❑ No

■ Have I included the following:
❑ Title page
❑ Table of contents
❑ Purpose
What it is:

_____

_____

❑ Hypothesis (for experiments)
What it is:

_____

_____

❑ Research
❑ Materials
Types:

_____

_____

_____

❑ Procedures
❑ Observations and results
❑ Conclusion
❑ Bibliography
What sources:

_____

_____

_____

❑ Acknowledgments
Who:

_____

_____

_____

Science Fair
Student Entry Form

To be returned before: _____

Student Exhibitor Information
(please print or type)

Name: _____
           Last   First

Age: _____ Grade: _____ Homeroom: _____

School: _____

Home Address:

_____

_____

Telephone: _____

Classification of Project (check one category)
- ❑ Biochemistry
- ❑ Botany
- ❑ Chemistry
- ❑ Computer Science
- ❑ Earth and Space Science
- ❑ Engineering
- ❑ Environmental Science
- ❑ Mathematics
- ❑ Medicine and Health
- ❑ Microbiology
- ❑ Physics
- ❑ Zoology
- ❑ Other: _____

Type of Project (check one category)
- ❑ Experiment
- ❑ Research
- ❑ Collection
- ❑ Apparatus
- ❑ Demonstration

Student Project Information

Title of project: _____

Brief description of project:

_____

_____

_____

_____

_____

_____

Electrical outlet required: ❏ Yes ❏ No

Display table required: ❏ Yes ❏ No

Special setup or arrangements required: ❏ Yes ❏ No
(If yes, please describe):

_____

_____

_____

_____

_____

_____

Parents' Agreement Form

TO HAVE YOUR ENTRY ACCEPTED FOR
EXHIBITION IN THE _____ SCIENCE FAIR
THE FOLLOWING STATEMENTS MUST BE SIGNED.

A. The project described above, which I plan to enter in the _____ Science Fair, is my own work and has been completed by me according to the rules of the science fair.

B. I understand that the project is entered at my own risk and that _____ is/are not responsible for loss or damage to my project or any of its parts.

C. I agree to leave my project in place until _____
                                                      date
and to remove it by no later than _____ on
                                              time
_____. If it is not removed by the designated time,
     date
I authorize that it be disposed of properly.

_____

Signature of Student

My son/daughter, or ward whose name appears at the top of this form and who has signed the statement above, has my permission to participate in the _____ Science Fair in accordance with its rules and regulations.

_____

Date

_____

Signature of Parent or Guardian

14 Questions & Answers

The following are questions about science fairs most often asked by teachers and administrators.

Q. Does every student have to participate?

A. The real value of a science fair is the opportunity to engage the entire classroom or school in worthwhile and exciting scientific discoveries. Promoting everyone's participation will help demonstrate the value of science in everyone's life. No student should be forced to participate, but all students should be encouraged.

Q. How can we "get the word out" about the science fair?

A. Take advantage of the local media. Send a press release to the local newspaper. Invite a reporter to do a brief story on the school's science program and the upcoming science fair. Encourage students to write their own "press releases" to take home and share with family members. Place notices and bulletins around the local community (in stores, on community bulletin boards, etc.). Print up inexpensive flyers and encourage students to distribute them throughout their neighborhood. Post information and articles about the science fair on school and community Web sites. Successful science fairs are promoted throughout the school and the local community.

Q. Our science fair has too many baking-soda volcanoes and solar system models made with table-tennis balls. What can we do?

A. A major reason for the proliferation of erupting volcanoes and bland solar system models is the lack of adequate ideas for students. By checking out all the resources and strategies in Chapter 6 (Helping Students Select a Topic), Chapter 7 (Idea Generators), and Chapter 8 (Suggestions for Projects), you can assist students in developing new and creative topics for exploration. Additionally, by encouraging students to plan six to twelve weeks in advance of the science fair, students have plenty of time to make thoughtful choices about potential projects. Volcanoes and solar system models (along with other common projects) are often selected because the decisions are made too close to the scheduled date of the science fair.

Q. How else can we get students to identify and select new and varied topics?

A. Encourage them to discuss their ideas or potential topics with you well in advance of the science fair. You may wish to initiate a plan in which students must have their project ideas "approved" by you or by a committee composed of teachers and/or other students by a specified date. Establishing a "topic committee" can ensure a broad spectrum of topics (instead of 437 volcanoes) across all the disciplines of science.

Q. How involved should parents get?

A. It seems as though some parents become part of the project itself, while other parents express no interest or commitment at all. Use the forms in this book to notify parents about a forthcoming science fair and to inform them about ways they can participate. Parental involvement is a valuable asset in any student's exploration of a scientific concept or idea. Carefully inform parents that they should serve as moderators and coaches, rather than as co-contributors with their children.

Q. Some students come from affluent families, while others come from families with limited economic resources. What can we do to ensure that our science fair does not give affluent students an unfair advantage?

A. First, it is important to emphasize to both students and parents that a science fair project *does not* require elaborate equipment or expensive supplies. Some of the most innovative and exciting projects are those created with common materials found around the home or in the local community. Parents need to understand that they are an intellectual resource for students, not a financial one. Second, you can minimize the influence of economic differences by providing time, space, and opportunity in the regular classroom schedule for students to accomplish some of the procedural steps in a science fair project. Plan regular and systematic chunks of time during the school day in which students can discuss ideas, share strategies, plan and record data collection, set up and design materials, write conclusions, and test project designs. In so doing, students will be able to plan and execute projects in an equally supportive environment.

Q. Does every project have to use the scientific method?

A. No. This book offers five different ways of designing and sharing a science fair project. These include experiments, demonstrations, research, collections, and apparatus. By emphasizing the variety of approaches to science fairs, you can help students understand the many ways to look at the world of science. While many students will want to generate a hypothesis and follow it through to a logical conclusion, it is not necessary for every project to follow this format. A variety of allowable formats will help ensure a multiplicity of projects and a broader range of student involvement.

Q. Can science fair projects use live organisms?

A. Discourage the display of live organisms in a science fair project. The safety or health of an organism cannot be guaranteed at a science fair. This is not to say that students couldn't investigate a question or develop a hypothesis about animals (e.g., "How does the fur on different animals help protect them from the elements?"). In such cases, much of the investigation will undoubtedly take place at home or in the community. Encourage students to take photographs and use the photos (rather than the actual animals) in the display. Initial planning and consultation with a teacher will help students design appropriate activities involving animals or plants.

Q. How do we handle students who copy information from encyclopedias, books, or Web sites and present it as part of their project?

A. One of the reasons students are prone to copy materials from another source and present it as their own is lack of sufficient planning. By using the planning forms in this book, you can help ensure that students have sufficient opportunities to explore a variety of resources. Also, you can promote original research by creating time in the classroom for students to access a variety of resources and to discussing that information with the teacher and/or other students. Careful monitoring over an extended period reduces the blatant use of the material of others. Also, you can reduce copying that results from the last-minute rush by helping students divide the project into manageable chunks. To do this, assign "due dates" for selected components of student projects (i.e., hypothesis, project design, written report, final display, and so on).

Q. How can I make sure that appropriate safety precautions are followed in the preparation and display of science fair projects?

A. Check out the section in Chapter 11 regarding safety guidelines. Be sure to share this information with students *before* any projects are started. Reproduce this list and distribute it to colleagues and parents, too.

Q. Where do we find judges for our science fair?

A. You can find science fair judges in local high schools, colleges, and businesses as well as in your students' homes. Invite high school science teachers to participate as judges. Call or write to one or more departments on the college or university campus and ask if any faculty members would be willing to judge a local science fair. In addition to the science departments (e.g., biology, chemistry, physics, engineering, astronomy), consider the education department, which undoubtedly will have professors who teach elementary or secondary methods courses in science.

Check around the local community at a factory, manufacturing plant, or commercial supplier for people with science backgrounds.

Parents are a terrific source for science fair judges. Seek volunteers among those parents who work in scientific fields, industry, or own their own business (e.g., appliance repair, hardware store, pet store). Do an informal interest inventory of parents early in the school year to discover their hobbies or interests (e.g., butterfly collector, rock hound, scuba diver, amateur astronomer).

Q. Does every science fair have to be judged?

A. No. The judging of a science fair is optional. A classroom science fair doesn't require formal judging since most of the students will be participating in one way or another. Some schoolwide science fairs do not need to be judged. For these events, you may wish to state that each entrant will receive a letter of participation rather than compete for ribbons or medals. Some of the best and most worthwhile fairs are those in which everyone's entry is celebrated regardless of its intent, depth, or sophistication.

Q. What is the most important factor in the success of a science fair?

A. Planning ahead. Take advantage of the timelines and planning materials included in this book. Actively involve students in this planning process. Make sure they understand that a science fair doesn't just happen, but rather that it is the result of several weeks (or months) of planning and preparation.

Q. Is there anything else that will help make the science fair successful?

A. Yes. Involve students as much as possible in all the dynamics of the science fair. Encourage them to ask their own questions, participate in the design of the fair, and engage in activities that provide them with a sense of "ownership" of the fair in general and their projects specifically. A science fair is not something done to students; rather, it achieves its greatest potency when it is something done *with* students.

15 Resources

These resources provide teachers, parents, and students with additional information on the preparation of projects as well as sources that can provide invaluable data for a selected topic or area of investigation. This chapter includes books and periodicals for students, teachers, and parents; sources for project supplies; and an extensive list of relevant Web sites.

Books for Students

Fredericks, Anthony D. (1996). *Exploring the Rainforest: Science Activities for Kids*. Golden, CO: Fulcrum Publishing.

Fredericks, Anthony D. (1998). *Exploring the Oceans: Science Activities for Kids*. Golden, CO: Fulcrum Publishing.

Fredericks, Anthony D. (2000). *Exploring the Universe: Science Activities for Kids*. Golden, CO: Fulcrum Publishing.

Fredericks, Anthony D. (2001). *Exploring the Human Body: Science Activities for Kids*. Golden, CO: Fulcrum Publishing.

Fredericks, Anthony D. (1995). *Simple Nature Experiments with Everyday Materials*. New York: Sterling.

Levine, Shar. (2000). *Quick-but-Great Science Fair Projects*. New York: Sterling.

Pearce, Q.L. (1997). *My First Science Fair Projects*. Los Angeles: Lowell House.

Smolinski, Jill. (1996). *50 Nifty Super Science Fair Projects*. Los Angeles: Lowell House.

Tocci, Salvadore. (1997). *How to Do a Science Fair Project*. New York: Franklin Watts.

Vecchione, Glen. (1997). *100 Amazing Make-It-Yourself Science Fair Projects*. New York: Sterling.

Books for Teachers and Parents

Bochinski, Julianne B. (1996). *The Complete Handbook of Science Fair Projects*. New York: Wiley.

Fredericks, Anthony D. (1997). *From Butterflies to Thunderbolts: Discovering Science with Books Kids Love*. Golden, CO: Fulcrum.

Fredericks, Anthony D. (1993). *Letters to Parents in Science*. Glenview, IL: Good Year Books.

Fredericks, Anthony D. (1998). *Science Adventures with Children's Literature: A Thematic Approach*. Englewood, CO: Teacher Ideas Press.

Fredericks, Anthony D. (2000). *Science Discoveries on the Net: An Integrated Approach*. Englewood, CO: Teacher Ideas Press.

Fredericks, Anthony D. (2001). *Investigating Natural Disasters Through Children's Literature: An Integrated Approach*. Englewood, CO: Teacher Ideas Press.

VanCleave, Janice. (1997). *Janice VanCleave's Guide to the Best Science Fair Projects*. New York: Wiley.

VanCleave, Janice. (2000). *Janice VanCleave's Guide to More of the Best Science Fair Projects*. New York: Wiley.

Periodicals for Students

3-2-1 Contact
(10 issues per year)
Children's Television Workshop
One Lincoln Plaza
New York, NY 10023

Audubon Adventures
(6 issues per year)
National Audubon Society
613 Riverville Road
Greenwich, CT 06830

Chickadee
(10 issues per year)
Young Naturalist Foundation
255 Great Arrow Avenue
Buffalo, NY 14207-3082

Dolphin Log
(6 issues per year)
Cousteau Society
8440 Santa Monica Boulevard
Los Angeles, CA 90069

Kids Discover
(Monthly)
Box 54206
Boulder, CO 90312-4206

KIND News
(9 issues per year)
The Humane Society of the U.S.
67 Salem Road
East Haddam, CT 06423-0362

National Geographic World
(Monthly)
National Geographic Society
17th and M Streets NW
Washington, DC 20036

Odyssey
(Monthly)
Kalmbach Publishing Co.
Box 1612
Waukesha, WI 53187

Otterwise
(Quarterly)
Box 1374
Portland, ME 04104-1374

Owl
(10 issues per year)
Young Naturalist Foundation
225 Great Arrow Avenue
Buffalo, NY 14207-3082

Planet Three
(10 issues per year)
Foundation Inc.
Box 52
Montgomery, VT 05470

Ranger Rick
(Monthly)
National Wildlife Federation
1400 16th Street NW
Washington, DC 20036-2266

Science Weekly
(Weekly during the school year)
2141 Industrial Parkway, #202
Silver Spring, MD 20904

Wonderscience
(Monthly)
American Chemical Society
1155 16th Street NW
Washington, DC 20036-4800

Your Big Backyard
(Monthly)
National Wildlife Federation
1400 16th Street NW
Washington, DC 20036-2266

Zoobooks
(Monthly)
Wildlife Education Ltd.
12233 Thatcher Court
Poway, CA 92064

Science Supplies

AIMS Education Foundation
1595 South Chestnut Avenue
Fresno, CA 93702
(888) SEE-AIMS
www.aimesedu.org

Bel-Art Products
6 Industrial Road
Pequannock, NJ 07440
(973) 694-0500
www.belart.com

Carolina Biological Supply Co.
2700 York Road
Burlington, NC 27216
(800) 334-5551
www.carolina.com

Delta Education Inc.
P.O. Box M
Nashua, NH 03061
(800) 258-1302
www.delta-ed.com

Edmund Scientific
101 East Gloucester Pike
Barrington, NJ 08007
(800) 728-6999
www.edsci.com

Educational Activities Inc.
P.O. Box 392
Freeport, NY 11520
(800) 645-3739
www.edact.com

Estes Industries/Hi-Flier
1295 H Street
Penrose, CO 81240
(303) 372-6565
www.estesrockets.com

Fisher Scientific Co.
4500 Turnberry Drive
Hanover Park, IL 60103
(800) 766-7000
www.fishersci.com

Frey Scientific
100 Paragon Parkway
Mansfield, OH 44903
(800) 225-FREY
www.freyscientific.com

Lawrence Hall of Science
University of California
Berkeley, CA 94720
(510) 642-5132
www.lhs.berkeley.edu

NASCO
901 Janesville Avenue
Fort Atkinson, WI 53538
(800) 558-9595
www.enasco.com

National Science Teachers Association
1840 Wilson Boulevard
Arlington, VA 22201
(703) 243-7100
www.nsta.org

National Geographic Society
1145 17th Street NW
Washington, DC 20036
(800) 647-5463
www.nationalgeographic.com

Wilkens-Anderson Co.
4525 West Division Street
Chicago, IL 60651
(773) 384-4433
www.waco-lab-supply.com

Other Materials

Showboard Inc.
P.O. Box 10656
Tampa, FL 33679
(800) 323-9189
www.showboard.com

This company manufactures standardized project display
boards ideal for use in science fairs. Showboard is sold at
many office and school supply stores or can be ordered
directly from the company.

Insights Visual Productions
P.O. Box 238644
Encinitas, CA 92023
(800) 942-0528
www.sciencevideos.com

This company produces a series of science videos that
focus on school science fairs. The videos and reproducible
print materials introduce teachers and students to the
requisites of well-designed projects and a well-developed
science fair.

Web Resources

The following Internet sites offer teachers, parents, and students a wealth of resources to ensure the success of science fair projects and the science fair itself.

Lesson Planning

http://www.cln.org/cln.html

The Community Learning Network is designed to help classroom teachers integrate technology into their classrooms. Its site has more than 5,400 annotated links to educational Web sites.

http://www.learner.org

A joint venture of the Annenberg Foundation and the Corporation for Public Broadcasting established this site to help schools and communities improve math and science education. Click on SAMI (Science and Math Initiatives) for a searchable database with easy access to curriculum resources, lesson plans and projects, and more.

http://www.afredericks.com

This site provides classroom teachers and school administrators with information about books, teacher resources, and training to improve elementary science education.

http://www.teachers.net/lessons

Take a lesson, leave a lesson at the Teachers Net Lesson Exchange. The lessons cover all subjects and grade levels, and include links to the teachers who posted them. This excellent Web site can energize the overall classroom curriculum.

http://www.pacificnet.net/~mandel

This site is a wonderful place to share ideas, concerns, and questions with educators from around the world. It provides lesson plans submitted by teachers in every curricular area. The material is updated weekly. The site also includes teaching tips for new and experienced teachers.

http://www.enc.org

The Eisenhower National Clearinghouse for Math and Science Education Web site has a searchable database with detailed descriptions of curriculum resources, articles from professional journals on science teaching, and favorite Internet sites for teachers.

http://www.teachnet.com/lesson/index.html

This site offers classroom teachers an array of lesson plans in every curricular area. This is an easy-to-use resource.

http://www.education-world.com

Education World provides lesson plans and other resources by curricular area as well as other articles of interest to teachers and administrators.

http://www.ceismc.gatech.edu/busyt

The Busy Teachers' Web site provides classroom teachers with links to a host of educational Web sites in a wide variety of curricular areas.

http://encarta.msn.com/schoolhouse/default.asp

This Microsoft Corporation Web site provides a treasure trove of lesson plans and a wide range of updated resources for every aspect of your classroom curriculum.

http://www.gsn.org

The Global Schoolhouse offers a diverse collection of Internet sites, videoconferencing opportunities, professional development activities, contests, and discussion lists for teachers and parents.

http://www.scholasticnetwork.com

The Scholastic Network offers an amazing variety of resources, including lessons and reproducibles and Web-based curriculum projects and activities.

General Science Fair Resources

http://www.isd77.k12.mn.us/resources/cf/steps.html

This site has short explanations of each part of a science fair and describes what makes a good project and how students can come up with their own science fair project.

http://faculty.washington.edu/chudler/fair.html

This site from a former science fair organizer and judge offers plenty of clear advice to assist students in generating, creating, and setting up a great science fair project.

http://www.capecod.net/~trowan/primer.html

The Science Fair Primer provides first-time science fair participants with concrete and easy-to-follow information on creating and developing projects from idea generation to final product.

http://sciencefairproject.virtualave.net/

Science Fair Project on the Web contains valuable information for students on the steps of a science fair project, the scientific method, and what judges look for in science fair projects. It also provides a list of teacher resources and links to sites that offer science fair ideas. This site is a great place to get started.

http://school.discovery.com/sciencefaircentral/

Among the many resources on school.discovery.com is a guide to science fair projects. The "handbook" features information from Janice VanCleave, a popular science author for teachers. Students can send her a question about their projects.

http://www.scifair.org/

The Ultimate Science Fair Resource offers a variety of resources and advice that will be useful to teachers and students in developing science fair projects.

http://sciencevideos.com/products/fair/science_fairs/scienc
e_fairs-1.html

This site provides links to many resources on getting
started, tips for teachers, generating ideas, sample projects,
and other information.

http://www.csun.edu/~lg48405

This site links to "virtual science fairs" from a number of
schools and science fair resources available worldwide. It
also links to research describing the benefits of posting
student research online.

http://atlas.ksc.nasa.gov/education/general/scifair.html

The scientists at the Kennedy Space Center have
participated in judging local school science fairs for many
years and have some great suggestions for conducting
student research projects.

http://setmms.tusd.k12.az.us/~jtindell/

A child-friendly site that offers ideas and resources for
getting started and using the scientific process to complete
a project.

Science Fair Project Ideas

http://www.halcyon.com/sciclub/cgi-pvt/scifair/guestbook.html

The Science Fair Idea Exchange offers an archive of science
fair projects and E-mail links to the students who created
them. Most items in the archive are short descriptions
posted by the students who did them. It has ideas for
biology, chemistry, physics, and psychology projects. It also
provides links to other science fair Web sites.

http://www.ascpl.lib.oh.us/scifair/

The Science Fair Project Index is a database of science
experiments-from simple to difficult-for students in grades
K–8. The index contains projects that have been published
in books since 1990. It provides the name of the
experiment and the book reference. Students can search by
general topics, by keywords, by materials or equipment
employed, or by the principle demonstrated. The index is
a project of the Akron-Summit (Ohio) County Public
Library.

http://www.stemnet.nf.ca/sciencefairs/

The New Science Fairs site lists ideas for science projects appropriate for grades 1–12. It provides links to other science fair resources and information about the Canada Wide Science Fair. It site is a project of the Eastern Newfoundland Science Fairs Council.

http://members.aol.com/ScienzFair/ideas.htm

ScienzFair™ Project Ideas lists ideas in 22 categories that students can use to develop into a science fair project. Some items are linked to additional Web references. New categories and ideas are added frequently.

http://www.scifair.org/

The Ultimate Science Fair Resource provides a wealth of information for students and parents, from finding an idea for a science project to tips for building displays. The Idea Bank lists general project ideas, and the Idea Board is a place where students post descriptions of projects they have done.

http://www.isd77.k12.mn.us/resources/cf/ideas.html

In addition to a list of science fair projects, Cyber-Fair offers tips to help students develop their own ideas. It provides a fill-in-the-blank idea generator and samples of student projects. The site was created by a science fair coordinator in the Mankato (Minn.) Area Schools.

http://www.exploratorium.edu/snacks/

For three years, nearly 100 teachers worked with Exploratorium staff members to create scaled-down versions of the museum's exhibits. The results were dozens of exciting "snacks"—miniature science exhibits that students can make using inexpensive, easily available materials.

http://youth.net/nsrc/webs.html#anchor453011

This site from the National Student Research Center has links to many "virtual science fairs"—schools that put their students' science fair projects on the Web. It also includes links to other sites that post student research.

Research

http://www.hpl.lib.tx.us/youth/science_fair_index.html

The Houston Public Library in cooperation with the Houston Independent School District created the Science Fair Page. It includes valuable information on defining a science fair topic, seeking information, using specific kinds of information, presenting the information, and evaluation.

http://www.isd77.k12.mn.us/resources/cf/SciProjIntro.html

This is an excellent site for students doing an experiment-based science fair project. There are links on this page to a more advanced guide and an example of an actual experiment-based project.

http://k12science.stevens-tech.edu/askanexpert.html

This site, developed by the Stevens Institute of Technology Center for Improved Engineering and Science Education, provides links for students to contact experts in a variety of subject areas.

http://www.ajkids.com

The Ask Jeeves for Kids Web site invites students to pose questions and provides kid-friendly Web sites through which they can obtain answers. This is a great resource for any area of the elementary curriculum.

http://ericir.syr.edu/Qa

AskERIC's question and answer service uses the diverse resources and expertise of the national Educational Resources Information Center (ERIC). The AskERIC staff responds to questions within two business days with ERIC database citations and publications, Internet resources, and referrals to other sources of information.

http://www.AskMe.com

Students and teachers can post questions on any subject and receive answers from self-appointed experts whose previous answers have been rated by other users.

http://www.askanexpert.com

This kid-friendly site connects students with hundreds of experts—ranging from astronauts to zookeepers—who have volunteered to answer student questions for free.

http://www.vrd.org/locator/subject.html

The AskA+ Locator links to more than 100 expert sites categorized by subject matter.

http://madsci.wustl.edu

The Mad Scientist Network bills itself as a "collective cranium" of scientists providing answers to questions posed by teachers, students, and the general public. It also contains archives of thousands of previously asked questions and answers.

Science Organizations and Competitions

http://physics1.usc.edu/~gould/ScienceFairs

At this site teachers and students can link to more than 100 science fairs from around the world. Specific science fairs at the national and international level, state science fairs, regional science fairs, and local science fairs are listed.

http://www.nsta.org
http://www.nsta.org/handbook/competitions.asp

The National Science Teachers Association offers a variety of resources for science teachers, including a position statement about science competitions.

http://www.macomb.k12.mi.us/ims/cr/science/so/nsoly/

The Science Olympiad is an international nonprofit organization devoted to improving the quality of science education, increasing student interest in science, and providing recognition for outstanding achievement in science education by both students and teachers. Science Olympiad competitions consist of a series of individual and team events that students prepare for during the year.

http://www.isef2001.org

This is the Web site for the 2001 International Science and Engineering Fair (ISEF), conducted by Science Services. It provides a list of rules for the high-level science fair competition. To obtain information for the current year, change the year in the Web address to the current year (e.g., www.isef2002.org, www.isef2003.org, etc.).

http://www.halcyon.com/sciclub

The Science Club offers activities that promote curiosity and the excitement of experimentation. Activities use common household materials. The site includes a Science Fair Idea Exchange.

Judging Science Fairs

http://www.geocities.com/Area51/Lair/7320/judgingstandards.html

The Maine Middle School Science Fair has adopted the same scoring procedure as that used by the National Science Fair. This one-page overview of judging criteria presents elements involved in the evaluation process.

http://www.isd77.k12.mn.us/resources/cf/rubric.htm

The Monroe School Science Fair Judging Form has specific point values for each item in the evaluation process. This form is a good prototype for other science fairs.

http://www.scifair.com/middle/judging.htm

The Massachusetts State Middle School Science Fair presents Topics for Consideration in Judging—a form that places point values for each of several different categories.

http://www.rain.org/~scifair/standards.html

The Santa Barbara (Calif.) Science Fair lists the Standards for Judging Exhibits on this site. Point values and specific categories are included.

http://ousd.k12.ca.us/~codypren/scifairjudge.html

This great article written by a science fair judge includes the trials and tribulations of judging along with some good advice for teachers. Also included is a student-generated rubric for evaluating science fair projects.